Modality and the Japanese Language

Michigan Monograph Series in Japanese Studies
Number 44

Center for Japanese Studies
The University of Michigan

Modality and the Japanese Language

Yuki Johnson

Center for Japanese Studies
The University of Michigan
Ann Arbor, 2003

Published by the Center for Japanese Studies,
The University of Michigan,
202 S. Thayer St., Ann Arbor, MI 48104-1608

Library of Congress Cataloging-in-Publication Data

Johnson, Yuki, 1956–
Modality and the Japanese language / Yuki Johnson.
p. cm. — (Michigan monograph series in Japanese studies ; 44)
Includes bibliographical references and index.
ISBN 1-929280-18-1 (cloth : alk. paper)
1. Japanese language—Modality. I. Title. II. Series.

PL629.M63 J64 2003
495.6'5—dc21

2002075088

Book design by City Desktop Productions

This book was set in Palatino.

This publication meets the ANSI/NISO Standards for Permanence of Paper for Publications and Documents in Libraries and Archives (Z39.48–1992).

Printed in the United States of America

Contents

Preface

"Any native speaker can teach language—badly."

The increasing level of sophistication in the field of language education—incorporating research from linguistics, language acquisition, and psychology—is appreciated by most people only in part, and it is all too common to encounter the opinion expressed in the former (pre-dash) portion of the above statement.

My career as a language educator started as a volunteer teaching Japanese pronunciation to Vietnamese children. I was still an undergraduate, and my main qualification for the job was not experience in language teaching—or in teaching at all—but as a radio announcer. My employers surely believed that if I could pronounce the mother tongue well, then I should be able to teach it well too. But my struggle began immediately. Not only did I strain to teach fairly simple matters, such as the proper pronunciation of the Japanese "R" sound, but I was at a loss in explaining grammatical characteristics of the language—for example, why *shinde-iru* only means "is dead" and not "is dying," whereas *sushi o tabete-iru* can mean either "I have eaten" or "I am eating."

Knowing a language is largely independent of knowing the rules underlying the language, which in turn is independent of knowing how to teach. To go beyond the first of these three, to defy the ignorance of the pre-dash portion of the above statement, and to teach well, one must study the language from the view of an outsider, adopting not the mind of the native speaker but that of the linguist and pedagogue.

My endeavors in linguistics were in large part motivated by the questions I received from students, questions that I could not answer immediately and systematically, and it is my desire to investigate the rules underlying language in a way that empowers language educators. This book was written in hopes of helping researchers, teachers, and students of Japanese further understand and appreciate the language from a linguistic viewpoint. The work began as part of my doctoral dissertation entitled "Japanese Modality: A Study of Modal Auxiliaries, Conditionals, and Aspectual Categories," completed in 1994, and soon after it was submitted, I began revisions of every aspect of the original work. Although this book project remains imperfect, and I feel the need for further revision already, I offer it nonetheless—a new segment on a long and ever-growing bamboo, with full expectation that the future will bring new segments to extend and strengthen the work.

This book attempts to answer a variety of problematic constructions from semantic viewpoints under the category of modality. My hope is that it will find audience with linguists, language educators, graduate students, and advanced-level Japanese learners. Research contribution to this volume spanned over a decade and was funded primarily by support from the University of Michigan and Center for Japanese Studies at the University of Michigan. I express my gratitude to those institutions and to the many individuals who helped to support this work.

I wish to express special thanks to my thesis advisor, Dr. Wesley Jacobsen, who guided me throughout my Ph.D. program and even after graduation. His pure devotion to linguistic research provided the best example of scholarly life and academic commitment. I also wish to thank the members of my original dissertation committee—Dr. Andrew Cohen and Dr. Elaine Tarone at the University of Minnesota, with whom I still have the pleasure to communicate from time to time. I also thank Dr. Takashi Masuoka, Dr. Yoshikazu Suzuki, and Dr. Atsuro Tsubomoto, who provided invaluable comments on the chapter concerning the *ba* conditional. In the time spent developing this book, I realized the importance of friends' encouragement and support. For that I am indebted to my colleagues, Dr. Esperanza Ramirez-Christensen and Dr. Hitomi Tonomura, and my friends, Dr. Brett Johnson, Ms. Mayumi Oka, and Ms. Shoko Emori. Many thanks to Mr. Bruce Willoughby for his editorial expertise and to Ms. Pamela Herzog for assistance with the index. And for their support, encouragement, and inspiration, I thank the two great physicists in my life—my father and my husband.

Finally, immeasurable gratitude goes to the person to whom I am most indebted for assistance with my research—Dr. Susumu Kuno. Like so many young linguists, I received countless insightful comments from Dr. Kuno that helped to guide me through my toughest research problems. With deepest respect, I thank him from the bottom of my heart.

Yuki Johnson
Spring 2001
Ann Arbor, Michigan

Key to Grammatical Abbreviations

ABL	ablative case (*kara*)
ACC	accusative case (*o*)
AUX	auxiliary (*hazu, ni-chigai-nai, daroo, kamoshire-nai, yoo, ~soo, rashii, beki*, etc.)
CAUS	causative affix (*~sase*)
CNT	contrastive marker (*wa*)
COM	comitative case (*to*)
COMP	sentential complementizer (*no, koto, toiu, tokoro*)
COND	conditional affix (*to, ~tara, ~eba, nara*)
COP	copula (*~da, ~desu*)
DAT	dative case (*ni*) (also indicates purpose)
DES	desiderative affix (*~tai*)
DIR	directional case (*e*)
GEN	genitive case (*no*)
GER	gerund affix (*~te*)
HON	honorific form (*~rare, o*-V *ni naru*)
IMP	imperative form
INF	infinitive form (∅, *~i, ~ku*)
INS	instrumental case (*de*)
ITER	iterative affix (*~te-iru*)
LOC	locative case (*ni, de, e*)
NEG	negative morpheme (*~nai*)
NPST	nonpast tense affix (*~u/ru*)
NOM	nominative case (*ga*)
PART	sentential particle (*yo, zo, ne*)

PASS	passive affix (~*rare*)
PST	past tense affix (~*ta*)
PERF	perfect affix (~*te-iru*)
POL	polite affix (~*masu*)
POT	potential affix (~*(r)are*, -*(r)e*)
PROG	progressive affix (~*te-iru*)
Q	question particle (*ka*)
QUOT	quotative form (*to*)
RES	resultative affix (~*te-aru*)
SPON	spontaneous affix (~*rare*)
STAT	stative affix (~*te-iru*)
TEMP	temporal particle (*ni*, *de*)
TOP	topic marker (*wa*)
VOC	vocative particle (*yo*)
VOL	volitional affix (~*yoo*)

Chapter 1

Introduction

This book attempts to refocus what has become a very broad definition of modality in the Japanese language (generally accepted as a speaker's psychological attitude) by building on the history of modality in order to create a new, practically oriented point of entry into its study. The work consists of six chapters: 1) an introduction; 2) a brief history of the study of modality; 3) an examination of modal auxiliaries; 4) an examination of aspectual categories; 5) a study of the interaction between modality and conditional sentences; and 6) a summary in terms of the relationship between modality, actual world situations, and stativity.

Chapter 2 traces the history of modality and addresses four main points: 1) the development of the study of modality in English; 2) the study of modality in Japanese; 3) establishment of a point of contact and common ground between English and Japanese; and 4) discussion of a different approach to the study of modality in the Japanese language which applies the notions of possibility and necessity to modal content.

Examination of the history of the study of modality reveals an unfortunate truth—that discourse in the subject often suffers from confusion and inconsistency—and raises an obvious question: is there a more coherent way to approach the study of modality in the Japanese language than has previously been attempted? "Modality" has become a loaded term concerning a speaker's psychological attitude; unfortunately, the term is often used under the assumption that everyone knows precisely to what it refers.

In the English language, the word modality finds its root in the Latin "modus" (manner) and has a research history spanning more than a century. Even though the term "modality" itself was not clearly defined, throughout this period the term "mood," represented by indicative, subjunctive, and imperative categories, was considered within the realm of traditional English grammar. Mood has often been restrictively defined as a category expressed by verbal morphology, "a syntactic, not a notional category, which is shown in the form of the verb" (Jespersen 1924, p. 313), or as "a grammatical category that is found in some but not all languages" (Lyons 1977, p. 848). On the other hand, modality has been described as "the semantic domain pertaining to elements of meaning that languages express" (Bybee 1995, p. 2). Modality may not necessarily be expressed within a verbal morphology in all languages; rather, it is a notional category that encompasses a wide range of concepts (obligative, potential, hypothetical, intentive, jussive, etc.) and grammatical categories (tense, aspect, negation, etc.). In addition, traditional modal logic has redounded to this study by classifying the categories of "epistemic" and "deontic," derived from the notions of "necessity" and "possibility." The definition of modality as the "opinion or attitude of the speaker" (Lyons 1977, p. 452) acknowledged by Palmer (1986, p. 2) is based on the observations of logical behavior of the English language in these epistemic and deontic categories. Recently many researchers have adopted this definition of modality, since it seems to encompass most of the concepts that represent mood. The concept of modality thus involves a number of categories and, as a result, is less clear than that of mood. This has spawned a lack of agreement on precisely what modality means.

In early twentieth century Japan, linguists were seeking to define sentence grammar more properly and were specifically looking at the relationship between subjects and predicates. The root of the study of modality in the Japanese language can be found in this early research. Later linguists would distinguish the propositional content of a sentence from its modal content. Propositional content was defined as some fact or description expressed by a speaker, and modal content as the part of the sentence which expressed a speaker's opinion or attitude toward the propositional content. This distinction communicates the notion of modality that was given voice by Fillmore (1968, p. 23) who also defined a sentence as "Sentence = Modality + Proposition," with "proposition" being the core part of a sentence—that is, which remains after removing the modal content (negation, tense, mood, and aspect, for example). The relationship between propositional content and modal content has thus been recognized as the study of modality in Japanese as well.

Compared to English, however, the Japanese typological distinction (subject, object, verb word order), poses both advantages and disadvantages in the study of modality. An advantage is that modal content can be distinguished from propositional content by the use of a tense marker as a syntactic divider. If close examination is given to each modal in terms of its syntactic behavior, however, this statement fails to provide a complete and rigorous rule. There are certainly exceptions in the way a modal follows a proposition in order to coordinate a sentence, such as with the modal auxiliary *soo* (e.g., in *Ame ga furi-soo da* "It looks like it is going to rain") following the stem form of the verb *furu* "to fall" and not the tense markers *ru* or *ta*. This syntactic irregularity should be considered a device that differentiates the meaning of the morpheme *soo* from its other meaning (hearsay). As Nitta (1989) states, the distinction between propositional content and modal content, in Japanese too, is based on a semantic-syntactic distinction, rather than a strictly syntactic one. However, due to such syntactic behavior, problems regarding ambiguous interpretation associated with English modal auxiliaries (such as "must" and "may") are avoided. In Japanese, two different forms are used to represent deontic modality (which expresses the degree of obligation, permission, or prohibition) and epistemic modality (which expresses the speaker's knowledge or belief). For example, "it must be hot" is expressed either by the deontic structure equivalent to "it has to be hot," (which is purely propositional) or by the epistemic structure equivalent to "I believe it is hot." This feature clearly illustrates that deontic modality is secondary to epistemic modality (since the latter is more closely associated with the speaker's attitude)—a concept that English modal auxiliaries cannot explicitly demonstrate.

A disadvantage is that all constituents of a sentence that follow a tense marker become the object of the study of modality, since these constituents are considered to be in the domain of modal content. Hence, modal content in Japanese involves a much wider variety of subcategories, such as auxiliary verbs and sentence final particles, and therefore corresponds neither to the idea of mood nor modality present in European languages.

This syntactic disadvantage has ensnared the study of Japanese modality in a chaotic tangle by spawning a multitude of subcategories. Lyons' definition of modality (as a speaker's opinion or attitude), after all, can be viewed in terms of any language phenomenon. Approaching modality from such a viewpoint implies that unlimited numbers of sentences and lexical items have to be considered before one can determine what sort of modality the sentence or the lexical item may or may not be able to express. In considering merely semantic views, one would have to sort through endless reams of data. Although a number of articles in Japanese concerning Japanese modality have been published, few fully explore the advantages exhibited by Japanese language structure. Furthermore, none provides sufficient analysis of the corresponding disadvantages; rather, these articles are concerned primarily with evincing problematic aspects of Japanese modality.

As a starting point for identifying a novel, practical entrance to the study of modality in Japanese, chapter 3 is devoted to an examination of Japanese modal auxiliaries that correspond to English modal auxiliaries. The concepts of epistemic and deontic modality, together with the notions of necessity and possibility as discussed in English language studies of modality are adopted as means of clarifying the functions of Japanese modal auxiliaries and their association with varying degrees of epistemic necessity and possibility.

The word "epistemic" is related to "epistemology," which is the philosophical study or theory of the nature and grounds of knowledge and belief (Hintikka 1969). In natural language, the term "epistemic" is concerned with a speaker's assumptions or assessment of possibilities, and it indicates a speaker's judgment along with the degree of a speaker's confidence or lack of confidence in the truth of the proposition expressed. Epistemic modality is strongly related not only to the possibility of the expressed proposition, but also to hypotheticality in the lowest degree of possibility of a proposition being true.

The word "deontic" is related to "deontology," which is the theory or study of moral obligation. Deontic modality is concerned with the necessity or possibility of acts performed by morally responsible agents (Lyons 1977), which in turn is represented by propositions that express the notion of permission, prohibition, or obligation. The common ground in these three concepts is that the speaker knows what will be brought about in some future state of affairs described by the proposition

By utilizing the results of previous research in both English and Japanese linguistics, Japanese modals are examined by applying to them the concepts of epistemic and deontic necessity and possibility in tests of both semantic and syntactic behavior. By means of such investigation, it is shown that the degree of modality is governed by the notion of epistemic possibility (namely, that modality is highest when a speaker's conviction is lowest). In practice, modal auxiliaries are chosen based on a speaker's knowledge and beliefs (which determine how strongly the speaker is convinced that a proposition is true), although, to varying degrees, they may also be chosen when the notions of possibility and necessity are considered. The higher (lower) the degree of speaker conviction that the proposition is true, the closer the proposition is to epistemic necessity (possibility). Similarly, the more (less) the speaker imposes the realization of the proposition, the closer the proposition is to deontic necessity (possibility). For both epistemic and deontic modality, regardless of the degree of speaker conviction, the expressed proposition is always concerned with a nonactualized situation. Therefore, both notions are in the domain of possible/nonactual worlds. Hence, modality can be thought of as being concerned with the expression of the degree of a speaker's conviction concerning a proposition's truth or realization in the form of possible/nonactual worlds. When a sentence is said to express modality, it is these "possible/nonactual worlds" to which "modality" specifically refers.

This observation is intriguing in that the notion of possible/nonactual worlds is conceptually very similar to psychological spaces in the theory of mental spaces proposed by Fauconnier (1994). He defined mental spaces as "constructs distinct from linguistic structures but built up in any discourse according to guidelines provided by the linguistic expressions" (p. 16). Under such a theory, epistemic modals play the role of "space builders" that allow the speaker to project an imaginary world, separate from an actual world, which is not necessarily true or realized. Such imaginary worlds also communicate "outside the territory of information" in Kamio's (1990) theory of territory of information in which there is a psychological distance between a speaker/interlocutor and the information described in a sentence. In Kamio's theory, the psychological distance is judged only by the notion of either closeness or remoteness to the speaker/interlocutor, and involves no gradation between these two notions. The examination of Japanese modal auxiliaries steps into Kamio's theory concerning the realm outside the territory of information and admits a different degree in the way a speaker perceives a situation. The common point in these theories is that neither possible/nonactual worlds nor imaginary worlds belong to the speaker him/herself internally; rather, they exist external to the speaker, who is therefore incapable of manipulating the status of the situation.

The definition of epistemic and deontic modality in English language studies is thus adopted as a way of clarifying the functions of Japanese modal auxiliaries and postulating that they can be approached in terms of possible/nonactual worlds. Ultimately, this examination points to the idea that the notion of possible/nonactual worlds is closely related to the concept of the indeterminability of a proposition—a concept which is inevitably expressed in hypothetical situations. This idea (which is derived from the syntactic and semantic features of the Japanese language) communicates the claim proposed by Jespersen (1924) that epistemic modality (containing no element of will) includes the subcategory of "hypothetical." Expressed by conditional sentences that have neither truth value nor factivity, hypothetical situations nonetheless express possible worlds in which one can determine neither the truth of the antecedent nor the consequent. Therefore, conditional sentences can be said to express modality. By defining modality through the expression of possible/nonactual world situations, modality becomes not merely a matter of modal content, but propositional content as well, and the study of modality thus proceeds to the examination of propositional content.

As mentioned above, propositional content is defined as some fact or description expressed by a speaker and is syntactically distinguished from modal content by use of a tense marker. Aspectual categories are certainly in the domain of the proposition. In chapter 4, the relationship between modality and aspectual categories is investigated, followed in chapter 5 by an examination of modality and conditionals.

In the Japanese language, aspectuality has been viewed as expressing various developmental stages of verbs, especially in terms of the dichotomy of *ru/te-iru* and *ta/te-i-ta.* The study of aspect in Japanese has long been an object of research in studies of verbs, yet the negation of verbs has somehow been left out of the aspectual category. The belief is that once an activity is negated it can no longer be broken down into developmental stages. Also, verbs are viewed as quantifiable in terms of how long it takes for an event to occur. This means that it is always expected that an event will occur at some point, and the negation that denies the occurrence of the event is therefore disregarded as an aspectual category. *Te-iru,* however, is used to express not only a progressive meaning, but also a resultative meaning. For example, *tabete-iru* "(I) am eating" can also express a resultative aspect, including the time of speech as well as the time that the speaker finished eating (in the past). In such a case, the interpretation of *tabete-iru* is "I have eaten." The function of *te-iru* here is to express a state resulting from an event, and this resultative aspect can be viewed as stativity, which itself does not change a situation. As noted by Comrie (1976), "for a state to remain the same requires no effort from inside or from outside the sentence" (p. 49). If we may draw analogy from far afield—it was contended by the physics of Aristotle that an object in motion will seek a state of rest, but that of Galileo argued that an object in motion will stay in motion unless acted upon by an outside force. Thus the physics of language—like that of the moon and of a pebble dropped from the Tower of Pisa—is Galilean: the state of the speaker having eaten will continue unless something happens to change that state.

Negative sentences possess this same feature because negation itself cancels the possibility of a change occurring as a matter of course. The fact is, a sentence is negated in Japanese by use of the negative suffix *nai,* which is morphologically the same as adjectives, in themselves regarded as stative predicates. Since *te-iru,* which is an aspectual category, expresses stativity, the negation indicated by *nai* (which also expresses stativity) should also be a candidate for inclusion in the aspectual category. This issue can be addressed by applying Dowty's (1979) predicate categorizations. Dowty divided predicates into nonchange and change predicates and labeled these categories "homogeneous" and "nonhomogeneous," respectively. Based on this categorization, both *te-iru* and *nai* are in the same "homogeneous" category. This means that *te-iru* and *nai* ought to possess some of the same characteristics. One may notice, however, that their functions are not completely identical in that the behavior of *nai* is often related to a particular negation's corresponding affirmative form. In chapter 4, *te-iru* and *nai* are examined in progressive constructions and *uchi ni* "while/before" constructions in order to clarify their similarities and differences within the aspectual category. The examination of *te-iru* and *nai* reveals the latter to belong in the aspectual category and further provides a basis for the study of its relationship with modality. This is done by investigating the features of internal negation and external negation structures, which in turn are concerned with the overall scope of negation.

Although negative sentences can have various meanings depending on context, in terms of modality, negative sentences are similar to *te-iru* sentences in that they lie outside the domain of modality subcategorized in this book. Simple negative sentences express the nonoccurrence of an event, but when such negative sentences accompany noun phrase (NP) complements, there are cases where negative *nai* represents nonactual world situations. The relationship between NP complements and predicate types are thus investigated within the structure of internal and external negation in order to discern whether or not those nonactual world situations are in the domain of modality.

Chapter 5 examines conditionals in terms of the degree to which they express hypotheticality. In Japanese there are three conditional forms: *to, ba,* and *tara,* which are variously equivalent to "when," "whenever," and "if" in English. Whether the Japanese conditional forms express "when," "whenever," or "if" depends on the nature of the consequent clause and on pragmatic judgments made by the speaker. It should be noted that these forms can be used to express situations which have already occurred, in spite of the fact that the word "conditional" usually refers to nonactualized situations. In this respect too, Japanese conditionals present certain properties that are quite distinct from their English counterparts. Thorough investigation is undertaken for each conditional in order to reveal the complexities of such from the perspective of modality. The examination goes beyond theories presented by previous researchers and demonstrates a new approach to the issue of conditionals. As a result, it is shown that among the three conditionals, *ba* conditional sentences express the highest degree of hypotheticality and, therefore, the highest degree of modality.

An interesting point in the study of conditionals was also found regarding the role of the aspectual form *te-iru* and negative suffix *nai*. Chapter 5 also includes an examination of *ba* conditional sentences containing aspectual categories and observes how they affect the interpretation of those conditional sentences. Of the two meanings expressed by *te-iru*, the resultative/stative meaning plays the greater role in the conveyance of hypotheticality, whereas progressive/iterative *te-iru* does not possess this function. In the examination of *nai* in *ba* sentences, it is discussed how and in what fashion the morphological feature of *nai* as an adjective and the status of *nai* as a suffix contribute the interpretation of *ba* sentences, and a deep correlation between stativity expressed by aspectual categories and modality is revealed. This also suggests that modality is concerned with static situations (rather than dynamic situations) where there is a distance between a speaker and a situation that cannot be controlled by the speaker.

The final chapter summarizes the conclusions in terms of the relationship between modality, possible world situations and stativity, specifically addressing how possible world situations and stativity are related to aspectual categories.

Example sentences used throughout this book were derived from three corpora: sentences created by the author for illustrative purposes; sentences drawn from examples employed by other researchers referenced as such; and sentences drawn from authentic materials, including private communications, Japanese television dramas and talk shows, Japanese novels, and Japanese newspaper articles.

The book also employs results gained from a questionnaire designed to obtain feedback from native speakers of Japanese. This questionnaire was distributed to forty native Japanese speakers between the ages of 30 and 40. The group included twenty women and twenty men, none of whom speak a second language. They were asked to answer the questions intuitively, and their responses were used to decide problematic points of commonly acceptable, practical language use.

Chapter 2

Evolution of the Study of Modality

2.1 THE DEVELOPMENT OF MODALITY IN WESTERN LANGUAGES

2.1.1 Modality in Logical/Philosophical Studies

The study of modality has been a theme of logical/philosophical studies for centuries, and its origin can be traced to ancient Greece, where "mood" was the subject of considerable debate, especially in the area of ontology. Since then, mood has been considered in both philosophical literature and logic. In the former, mood has been enumerated into grammatical categories (verbal forms) such as indicative, subjunctive, imperative, optative, infinitive, and so forth and said to express a "certain attitude of the mind of the speaker toward the contents of the sentence" (Jespersen 1924, p. 313).

Mood as a grammatical category, however, is exclusive to the Latin language where verbs have a diversity of endings. In English, auxiliary verbs commonly represent a particular shade of meaning expressed in moods. Based on the conceivable meaning of moods in English, Jespersen (1924), for example, proposed a variety of notions that are classified under two subcategories: moods "containing an element of will" and those "containing no element of will." The former category includes sentence types such as imperative, obligative, permissive, optative, desiderative, and intentional, whereas the latter category includes sentence types such as necessitative, potential, conditional, presumptive, and apodictive.

Much later, Rescher (1968) introduced many other notions regarding the categories of mood, such as epistemic, temporal, boulomaic, deontic, evaluative, and causal, in order to subcategorize the concepts of modality. As Jespersen (1924) pointed out, "There are many 'moods' if once one leaves the safe ground of verbal forms actually found in a language" (p. 321). It is unavoidable to face difficulties when one tries to transform the Latin modal system into English, of course, since the structure of English is quite different from that of Latin. Furthermore, even within the Latin language itself, where mood is a formally grammaticalized syntactic category, the syntactic behavior of the language and the semantics cannot be completely separated; hence, mood cannot be viewed only from a single perspective, especially when a subject is concerned with the mind of a speaker. Thus moods in Latin are described by a multitude of sub-categories which in English are approached from a semantic viewpoint. This semantic viewpoint is common to the recent study of modality and is largely responsible for its presently chaotic state.

In logical studies, the notions of necessity and possibility were the central categories of discussion. According to Jespersen (1924, p. 319), Wolff attempted to place moods into the categories of possibility, necessity, and contingency, whereas Kant employed categories of possibility, necessity, and existence under the class of "modality." These notions then became the object of examination in "symbolic modal logic." It is in these logical studies dealing with modal concepts that the inception of the recent study of modality in natural language can clearly be recognized.

More recently, von Wright (1951) introduced four groups of modal concepts: alethic[1] modes, that is, the proposition is necessarily true, possibly true, or contingently true; epistemic modes,[2] that is, the proposition is known to be true or known to be false; deontic modes, that is, the proposition is concerned with obligation, permission, or

1. The word alethic originates in the Greek word alethic, which means truth.

2. The term epistemic originates in the Greek episteme, which means "knowledge." Hintikka (1969) defined epistemic modality as what we know and believe.

prohibition; and existential modes that is, the proposition is concerned with universality, existence, or emptiness of properties or classes). Of the four groups the alethic category covers most of what is traditionally dealt with in modal logic. Necessity and possibility have thus traditionally been the central notions of modal logic and are strictly within the domain of alethic modes. For example, the following notations represent the relationship between necessity and possibility. (*L* represents necessity, *M*, possibility, *p*, proposition, ~, negative proposition, and ≡, logical equation.)

(1) a. $Lp \equiv {\sim}M{\sim}p$
What is necessarily true is what is not possibly not true.

c. ${\sim}Lp \equiv M{\sim}p$
What is not necessarily true is what is possibly not true.

d. $L{\sim}p \equiv {\sim}Mp$
What is necessarily not true is what is not possibly true.

b. ${\sim}L{\sim}p \equiv Mp$
What is not necessarily not true is what is necessarily not true.

Notation (1)a, for example, can be represented by the following situation: if John is in his office every Monday between 2:00 and 3:00 and if it is now 2:30 on a Monday, then he must be in his office. Therefore, the logical interpretation is that "it is not possible that John is not in his office."

Human perception and judgment were factored into the notions of necessity and possibility and were combined with the ethical notions of "morality" and "law." As introduced by von Wright, their meaning was thus extended to "obligation," "prohibition," and "permission." Notions related to these three concepts fall under the category of deontic modality, and the relationship among them can best be seen by examining how their negative forms interact: if an act is not permitted, then it is prohibited; if the negation of an act is prohibited, then the act is obligatory; if an act and its negation are both permitted, the act is indifferent. The basic use of deontic modes in natural language, then, is to urge that a situation in question be realized. Therefore, deontic modes themselves express nonactualized situations.

Even the notions of necessity and possibility are examined in terms of logical truth. When a proposition is said to be contingently true, it implies that "there are other possible worlds, or states of the world, of which it is , or might be, false" (Lyons 1977, p. 787). As a way of taking into account nonactualized situations, modal logic eventually came to be calculated in terms of so-called "possible worlds," which are a means to conceive the possibility of an alternative way in which the world might have been. Lewis (1973) views "possible worlds"[3] in the following manner:

> It is uncontroversially true that things might have been otherwise than they are. I believe, and so do you, that things could have been different in countless ways. But what does this mean? Ordinary language permits the paraphrase: there are many ways things could have been besides the way they actually are. On the face of it, this sentence is an existential quantification. It says that there exist many entities of a certain description, to wit, "ways things could have been." I believe permissible paraphrases of what I believe; taking the paraphrase at its face value, I therefore believe in the existence of entities which might be called "ways things could have been." I prefer to call them "possible worlds." (p. 84)

Hintikka (1969) defined possible worlds as a set of formulae that can be thought of as a partial description of possible states of affairs. This means that what is possibly true in some state of affairs *W* (where *W* is the set of all possible circumstances) is true in at least one of its alternatives, that is, if a proposition *p* is nonempty, then there is at least one circumstance in which it is true. On the other hand, that which is necessarily true in *W* is true in all of its alternatives, that is, there is no circumstance in which it is false. These formulae are then calculated in the possible world semantics known as intentional logic (Chierchia and McConnell-Ginet 1992, pp. 213–222).

Possible worlds can be summarized as "not concrete objects or situations, but abstract objects whose existence is inferred or abstracted from the activities of rational agents" (Stalnaker 1984, p. 50). They describe a possible but nonactual state of affairs and represent "what would be the case if some events had different outcomes from those

3. The formal apparatus of possible worlds was, in fact, first introduced by Kripke (1959) as a tool for investigating the semantic properties of certain formal systems.

that they in fact have" (Chierchia and McConnell-Ginet 1992, p. 206). Possible worlds are further examined in temporal discourse examining the states of a particular world at different points in time, that is, the examination of possible worlds includes the category of tense.

Necessity and possibility themselves were first studied in logic via the category of alethic modes, a category that deals specifically with the truth of propositions. Then as linguists sought to incorporate notions of human belief and knowledge, there was a shift in emphasis, and modality came to be studied under the domain of epistemic modes. Under this category, the conception of modality has exploded to include the use of "possible world" concepts as a way of furthering its definition. Recognizing that these notions could not be handled strictly within the domain of alethic modes, and that they also include epistemic modes, logicians created possible world semantics, where notions of necessity and possibility are conceived as the elements of intentional logic. Since necessity and possibility are concerned with human "knowledge and belief" in ordinary usage as pointed out by Hintikka (1969), these notions were eventually applied to the study of natural language.

2.1.2 Modality in Natural Language

In English, moods are studied under a grammaticalized category of auxiliary verbs—the so-called "modal auxiliaries" such as "may," "must," "should," "can," and so forth. This terminology was first used by Poutsma (1926) to designate the expression of mood by such auxiliaries, which are also concerned with the notions of necessity and possibility.

These notions cannot always be rigorously interpreted, however, since they involve human judgments. When necessity is viewed in terms of epistemic modes, the implications of natural language can seldom be strictly defined, since the modal "must" carries both an obligatory meaning and a high degree of speaker conviction. It is unavoidable that the approach to the notions of necessity and possibility has been by way of the notion of "epistemic," which involves the world in which we live, wherein we can create nonactual situations, conditions, as well as inanimate and animate things. Human imagination, judgment, and creativity are much broader concepts than can be handled in the confines of symbolic logic.

As was seen in symbolic logic, what is necessarily true is what is not possibly not true ($\sim Lp \equiv M\sim p$).[4] In English, the modal "must" often serves to express the notion of necessity. "Must," however, can be interpreted as either "have to" (implying obligation) or "something is expected" (implying a high possibility of the proposition's truth) depending on context. In everyday speech, "must" is frequently used in an epistemic sense. An example of this is found in Karttunen (1972, p. 15):

(2) If Bill has a diamond ring, he must have stolen it from someone.

Karttunen claims that "must" in sentence (2) does not express logical necessity regardless of whether it is predicated in the conditional as a whole or in the consequent alone. Modals like "must" should be dealt with in the domain of epistemic modality, and should not be confused with their counterparts in symbolic modal logic.

In epistemic modality the notions of necessity and possibility are approached with regard to a speaker's belief and knowledge, which are themselves concerned with a speaker's judgment toward a proposition. Epistemic necessity, therefore, is interpreted as: based on what the speaker knows, it is necessary that the proposition is true. This also communicates that: based on what the speaker knows or what s/he (confidently) infers, the proposition is true. Epistemic possibility is interpreted as: based on what the speaker knows, it is possible that the proposition is true. In other words, the speaker has a low degree of conviction as to the truth of the proposition.

Branching off from symbolic logic, it has become prevalent to examine modality as a subfield of syntax/semantics and pragmatics. Syntactic significance can first be found in Chomsky (1965) who introduced modality in the phrase structure rule, "Aux → Tense (M) (Aspect)." This "M" is a category which includes auxiliary verbs such as "will," "shall," "may," and so forth, and is understood here in terms of "mood." Following Chomsky, Fillmore (1968) proposed that the category he labeled "proposition" was one of the most basic constituents of a sentence, illustrated in the rule, "Sentence = Modality + Proposition." There "proposition" is defined as the core part of a sentence: that which remains after removing the modal content (negation,

4. Example (1)a.

tense, mood,[5] and aspect, for example). This supposes that a distinction can be made in a sentence between modality and proposition, although unfortunately Fillmore does not provide a clear definition of modality.

A semantic interpretation of modality was distinguished by Palmer's quotation of Lyons' "opinion or attitude of the speaker." This definition of modality seems to be widely accepted among linguists, especially in those languages which do not have equivalents to English auxiliaries. Palmer (1986) adopted Lyons' definition literally and in doing so broadened the field of the study of modality.

Palmer (1986) examined various languages and categorized the various types of modality he found. The difference between "mood" and "modality" he defines as follows:

> "Mood" is traditionally restricted to a category expressed in verbal morphology—'a syntactic, not a notional category, which is shown in the form of the verb' (Jespersen 1924, p. 373). While "modality" is not expressed in all languages within the verbal morphology, it may be expressed by modal verbs or particles which may well be quite separate from the verb. (p. 21)

According to Palmer, modality can be expressed by categories other than the verb. The broad definition of modality he proposed thus goes well beyond most traditional treatments of English modals. For example, he placed "evidentials," linguistic forms which indicate secondhand information, in the category of epistemic modality, and he labeled them "judgments." He also broadened the deontic category by including "imperatives," "volitives" such as those expressed by the verbs "wish" and "hope," and "evaluatives" such as the expressions "good job" and "bad job." Palmer further claimed that modality could be viewed not only in terms of epistemic and deontic categories such as possibility and necessity, but also in terms of subjectivity and factuality. Under such circumstances, he included declarative, complement, and oblique clauses, as well as tense, aspect, negation, and gender in his examination. Based on this definition, however, all language phenomena can be viewed in terms of modality. Even though

5. For Fillmore (1968), indicative, subjunctive, and imperative are all kinds of mood, and he uses the word "modality" to indicate a much wider and more abstract category.

it is true that his study is suggestive in many ways, Palmer's polysemantic approach paints a chaotic picture of the study of modality. The influence of his study can be observed in the use of a variety of terminology, such as "root (nonepistemic modality)" (Coate 1983) and "realis/irrealis" (Bybee 1995) that have been introduced in recent studies of modality.

What becomes evident in examining the history of modality is that logical formalizations of modality are concerned with precise categories and do not succeed in solving the problem of indeterminacy in natural language. The study of modality now has become independent from that of modal logic and involves various language phenomena. The polysemantic method emphasizes this approach by dealing with many different categories that had been regarded as discrete: evidentials, declarative statements, conditionals, and so forth. Neither method may be fully satisfactory for all languages, but it should be recognized that by proposing a multitude of theories researchers have created inroads toward a new conception of modality.

2.2 PREVIOUS STUDIES OF JAPANESE MODALITY

2.2.1 Japanese Auxiliaries

In English, modality has been studied by examining "moods," which are intimately related to auxiliary verbs such as "shall," "will," "may," "must," "can," "be," "have," "do" and so forth. Auxiliary verbs are all morphologically independent due to inflectional features of the English language and are used in conjunction with another verb to express "person, number, mood, or tense" (Merriam Webster's New Collegiate Dictionary, tenth edition). As the name "auxiliary verb" implies, they are "verbs" which support a main verb.[6] For example, both "might" and "have" are auxiliaries in "He might have gone," where "might" belongs to a category of modal verbs used with other verbs or in elliptical sentences where a verb is understood, and "have" is an auxiliary related to the past

6. The question of whether or not these English auxiliary verbs are main verbs is a point of controversy among English researchers. For example, Ross (1976) claims that they are main verbs and Palmer (1979) opposes that premise.

participle (distinguished from "have" which functions as a full verb such as in "I have a car"). "Be" and "do" can also be used as main verbs, as in "I am a student" (as opposed to "I am going") and "I do my job" (as opposed to "I do like strawberries"). In these sentences, the "am" in "I am a student" and the "do" in "I do my job" are not considered auxiliary verbs; rather, the term "auxiliary" describes the function of the verb and does not designate the part of speech in the same way as the terms "adjective," "noun," "verb," and so forth. Auxiliary verbs therefore belong to a small category that syntactically accompanies other verbs and expresses particular shades of meaning. In English linguistic research the term "auxiliaries" is often substituted for "auxiliary verbs" as a way of emphasizing which function takes precedence in a particular example (e.g., modal auxiliaries).

The Japanese equivalent to auxiliary verbs is *jodooshi* (auxiliary, hereafter) which was first introduced as a part of speech by Ootsuki in 1904. According to Ootsuki, an auxiliary is a constituent which follows immediately after the stem form of a verb, thereby changing the form and meaning of the verb. While auxiliaries are similar to English auxiliary verbs in terms of their function and meaning, due to the agglutinative feature of Japanese, most auxiliaries are not morphologically independent, but rather are attached to the main verb in a strictly defined order. This morphological feature creates a completely different picture from how English auxiliaries are viewed. In modern Japanese, auxiliaries can primarily be classified into five types according to the form of the preceding word: i) those which are suffixed to the stem of a verb; ii) those which are suffixed to the so-called *te* form of a verb;[7] iii) those that are suffixed to the stem/gerundive form of a verb (indicating meanings such as potential, passive, and causative; iv) those which indicate forms such as negation and tense (*ru, ta,* and *da*); and v) those which follow the proposition (after the tense markers *ru* or *ta*). The following example sentences represent each type listed above, respectively:

(3) i) Nihongo no kotoba wa oboe-nikui.
Japanese GEN words TOP remember-AUX
"Japanese words are difficult to memorize."

7. These auxiliaries can also function as independent verbs, but when they become constituents of a predicate, their original meaning become less distinctive.

(4) ii)

Tomodachi	ga	kuru	node,	heya	o	soojishite-oi-ta.
friend	NOM	come	since	room	ACC	clean-AUX-PST

"Since my friends are coming, I cleaned the room (in advance)."

(5) iii)

Jon	wa	sushi	o	tabe-sase-rare-ru.
John	TOP	sushi	ACC	eat-CAUS-PASS-NPST

"John is going to be forced to eat sushi."

(6) iv)

Taroo	wa	it-ta	ga,	Hanako	wa	ik-anakat-ta.
Taro	CONT	go-PST	but	Hanako	CONT	go-NEG-PST

"Taro went, but Hanako did not go."

(7) v)

Honda san	wa	kinoo	koko	e	ko-nakat-ta	kamo shire-nai.
Mr. Honda	TOP	yesterday	here	LOC	come-NEG-PST	AUX

"Mr. Honda may not have come here yesterday."

The auxiliary *nikui* "difficult" in sentence (3) is an adjective in its original use and, together with the gerundive form of the verb *oboe(ru)* "to memorize," constitutes the predicate of the sentence. In a similar manner, the past tense form of the verb *oku* "to put" is attached to the *te* form of the verb *soojisuru* "to clean," and together they constitute a predicate *shite-oi-ta* meaning "cleaned the room for later use." In sentence (5), the gerundive form of the main verb *tabe(ru)* "to eat" occurs with the auxiliaries *-sase(ru)-* "causative," and *-rareru* "passive," which are attached in a particular order to constitute the predicate of the sentence. As discussed below, the suffixes *ta* and *nai* (in *nakat-ta*) in sentence (6) were once considered auxiliaries in Japanese. The auxiliary *kamoshire-nai* in sentence (7), on the other hand, exhibits a slightly different characteristic compared to sentences (3)–(6). Whereas the auxiliaries in sentences (3)–(6) are agglutinated to the main verbs in a rigid order to alter the meaning of the predicate, *kamoshire-nai* does not suggest such strong morphological agglutination. Rather than forming a constituent with the preceding tensed verb form, *kamoshire-nai* forms a constituent together with the entire sentence preceding it. This observation is supported by the following examples:

(8) a. Nihongo no kotoba wa oboe-yasui-desu ka
Japanese GEN words TOP remember-AUX-COP Q
"Are Japanese words easy to memorize?"

* b. Hai, yasui-desu.
Yes cheap-COP
"Yes it is cheap."

(9) a. Jon wa sushi o tabe-sase-rare-masu ka
John TOP sushi ACC eat-CAUS-PASS-NPST Q
"Is John going to be forced to eat sushi?"

?? b. Hai, sase-rare-ru-n-desu.
Yes CAU-PASS-NPST-COMP-COP
"Yes (he) will be forced to (eat)."

c. Sensei ni shik-(r)are-ta-no?
teacher by scold-PASS-PST-COMP
"Were you scolded by your teacher?"

* d. Un, rare-ta-no.
yes PASS-PST-COMP
"Yes, I was."

(10) a. Nihon e kaeru kamoshire-nai-no?
Japan LOC return AUX-COP
"You might go to back to Japan?"

b. Un. kamoshire-nai.
Yes maybe
"Yeah. Maybe."

Yasui in sentence (8)a can create the meaning "easy to ~" only when it is suffixed to a verb; therefore, when *yasui* is used to reply to a question without the verb, as in (8)b, it functions no longer as an auxiliary, but as an adjective. The answer in (8)b is therefore inappropriate. Similarly in sentences (9)a and (9)b, although ellipsis of the verb in the response may occur in casual conversation since the meaning can be understood from context, a predicate consisting only of suffixes is not well-formed; it is syntactically an obscure sentence. The obscureness becomes greater when a consonant verb is used in the passive construction, as seen in sentence (9)d. On the other hand, the degree of acceptability of *kamoshire-nai* in (10)b is much higher than the responses in examples (8)b, (9)b, and (9)d. *Kamoshire-nai* is considered a semi-independent constituent used to give a speaker's judgment toward the preceding sentence. Therefore, there is a significant difference in terms of the degree of syntactic agglutination between the auxiliaries.

Due to the syntactic difference and semantic independence mentioned above, auxiliaries have been examined by various linguists primarily regarding whether or not they constitute a singular category that can be labeled as a part of speech. This debate can be traced back to a number of studies in classical Japanese, which has a morphology somewhat different from modern Japanese. Matsushita (1930), Yamada (1951), and Suzuki (1972) sought to establish terminologies that represent different features of auxiliaries based on their syntactic characteristics.[8] These researchers thought that auxiliaries do not constitute an independent part of speech of their own; they assumed that only free-standing morphemes (i.e., words) could be considered distinct parts of speech.

Taking an opposite position, Hashimoto (1948) postulated that auxiliaries could be studied as parts of speech. He defined auxiliaries as the forms that do not change the relationship between the original phrase and the phrase that results when these forms are attached. Then, he divided them into two categories: those attached to nouns and adjectives and those attached to verbs. This classification of auxiliaries had a large influence on later research, especially on how phrases are viewed, and suggested that sentences can be divided into two distinct segments. The following diagrams illustrate this idea:

8. For a more detailed history of auxiliaries, refer to Kondoh (1989) and Suzuki (1990), etc.

(11)a.

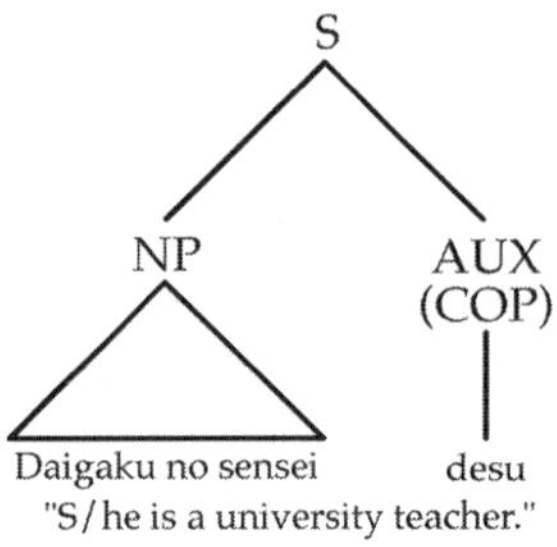

(11)b.

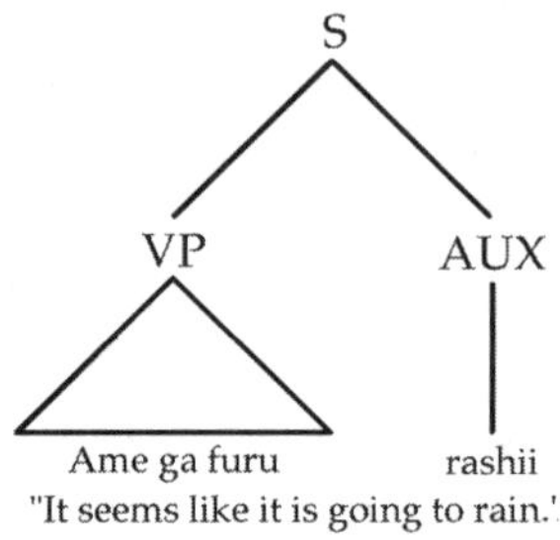

For Hashimoto, the copula *da* is considered a part of speech even though it is a dependent constituent, and its syntactic behavior is quite different from that of *rashii* in (11)b. According to Hashimoto's premise, however, suffixes such as causative, passive, potential, and spontaneous forms do change the relationship between the original phrase and can no longer be considered auxiliaries. Hashimoto set these constituents aside as *setsubi-go* "suffixes."

Among the researchers who further pursued Hashimoto's view of auxiliaries is Tokieda (1950). Acknowledging Hashimoto's critical identification of sentence structure, Tokieda confirmed that Japanese sentences can be divided into two basic domains,[9] termed *Shi* and *Ji*—the former expressing objective notions, and the latter conveying a speaker's subjective attitude toward the objective notions expressed in the *Shi* domain. The dichotomy is illustrated in the following diagram:[10]

(12)

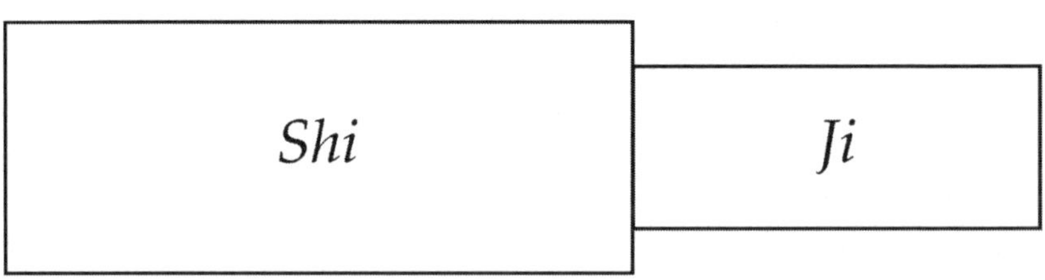

9. The difference between Hashimoto's view and Tokieda's view involves their treatment of particles, which is beyond the scope of this book.
10. This illustration is called *Ireko-gata koozoo* and describes the sentence structure as *Shi* being enveloped in *Ji* (Tokieda 1955, p. 278).

In Tokieda's view of sentence structure, dependent constituents such as the negative morpheme, *nai,* the past tense marker, *ta,* and the copula *da,* are all auxiliaries and are included in the *Ji* domain, while equally dependent constituents such as causatives and passives, are in the domain of *shi.* This determination is based on the semantic features of the constituents,[11] and the *Shi/Ji* distinction is often referred to as the origin of the study of modality in Japanese—namely, in the dichotomy of propositional and modal content.

2.2.2 Propositional Content and Modal Content

Tokieda's view thus had a significant influence over subsequent research on Japanese sentence structure and gave rise to a major debate called *Chinjutsu ronsoo*[12] that is concerned with the border between the *Shi* and *Ji* segments of a sentence. In modern Japanese, most researchers have come to an agreement that Japanese sentences can be divided into propositional content and modal content using a tense marker as a divider—viz, constituents that follow either *ru* or *ta* are in the domain of modal content.[13] Propositional content expresses an objective statement while modal content expresses a speaker's subjective judgment or attitude toward the proposition. Since then, the study of modality experienced a flood of new terminology aimed at defining modality in Japanese. Mikami (1963), proposed the term *Muudo* to describe modal content, a combination of the English "mood" and Yamada's *Chinjutsu* (1951). Teramura (1982) used the term *Koto* to refer to the propositional content and the term

11. For example, the negative morpheme can be used to express negative judgment, whereas the past tense and perfective markers can be used to express judgment of reminiscence and judgment of confirmation, respectively. The copula, on the other hand, conveys affirmative judgment.
12. *Chinjutsu ronsoo* means "predication debate," which was originally proposed by Yamada (1951). Yamada's *chinjutsu* is the equivalent of Tokieda's *ji,* which expresses the speaker's subjective judgment. Yamada claimed that verbs, adjectives, adjectival nouns, *jodooshi* and sentence final particles are not altogether dissimilar, and taken together constitute a predicate. A more detailed explanation of *Chinjutsu ronsoo* can be found in Ookubo (1968).
13. There are problems with this assumption, however. For example, *in Kare wa haiyuu rashii* "He seems to be an actor," *rashii,* which represents the modal content of the sentence, immediately follows the noun phrase *haiyuu* with no tense marker in between. Although the theory seems reasonable for most of the modals, the distinction between the propositional component and the modal content of a sentence cannot be made purely on a syntactic basis. A semantic viewpoint needs to be incorporated in order to support the theory, an issue which will be discussed at the beginning of chapter 3.

Muudo to refer to a speaker's subjective judgment or attitude. Nakau (1979) used the terms *Meedai* and *Modality,* and Nitta (1989) employed *Genpyoo jitai* and *Genpyoo taido* to express essentially the same distinction. All these terms are based on a dichotomy of propositional content and modal content. This distinction was born from the serpentine history of Japanese linguistic studies and appears to be endemic to the Japanese language.

According to this dichotomy, constituents that had been traditionally treated as auxiliaries fall in two large categories: suffixes and auxiliaries. Suffixes are categorized according to their dependent morphological features as: causative, passive, potential, and spontaneous forms as well as tense markers, negative morpheme, and the copula *da.* On the other hand, auxiliaries are categorized with respect to whether they belong to the propositional content or the modal content of the sentence. The former pertains to auxiliary adjectives and verbs that are attached to various verb forms in order to explicate meaning [e.g., *oboe-nikui* "difficult to memorize" in example (3) and *soojishite-oi-ta* "cleaned the room in advance" in example (4)]. The latter pertains to auxiliaries in modal content and are used to express a speaker's psychological attitude (mood) toward a proposition and are termed *Hoo* (mood) *Jodooshi* (auxiliaries)—*Hoo-jodooshi* (modal auxiliary—hereafter referred to as "modal"). Teramura (1984), for example, classified Japanese modals into two categories: modals that express conjectural mood and modals that express explanatory mood.[14] Although the classification established by Teramura does not precisely correspond to that of English modals, and the syntactic behavior of Japanese modals is not the same for all modals, the term *Hoo-jodooshi* reflects the equivalence of "auxiliaries that express mood," and it should be recognized as a reasonable nomenclature.

The following diagram illustrates the dichotomy of propositional content and modal content represented by sentence (13), followed by a chart (14) which exhibits the types of auxiliaries used in both propositional content and modal content:

14. The negative form of explanatory modals such as *-no-da* and *wake-da* are also used to express external negation structures and suggest that external negation is in the domain of modality. This subject is discussed in chapter 4.

(13)

Taroo	wa	kusuri	o	nom-ase-rare-nakat-ta	rashii.
Taro	TOP	medicine	ACC	take-CAUS-PASS-NEG-PST	AUX

"It seems that Taro was not forced to take the medicine."

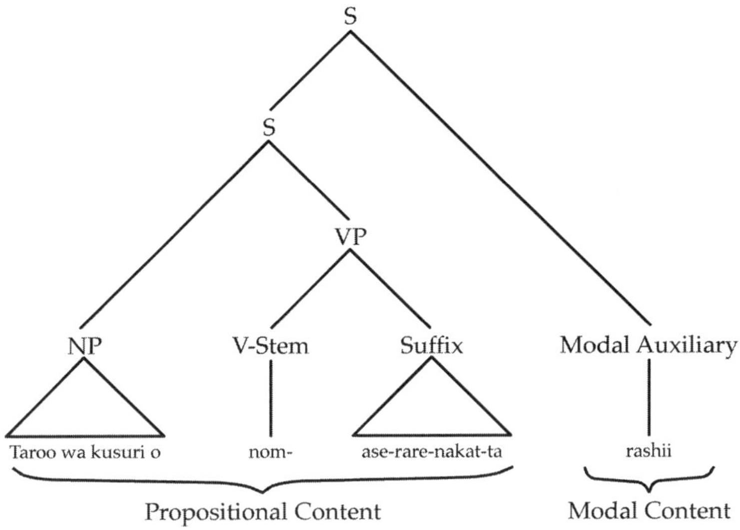

(14)

Propositional Content	Suffixes	*-ru, -ta, -nai, -da* passive, honorific, potential, causative, etc.
	Auxiliaries	Verb stem + Adj: e.g., *-yasui, -tai, -gatai,* etc. Verb gerund+ Aux V: e.g., *-teoku, -temiru,* etc.
Modal Content	Modal auxiliaries (Modals)	*wake, mono, koto, tokoro, no(da), hazu, ni chigai-nai, daroo, kamoshire-nai, yoo, mitai, soo, rashii, beki, mai*

Although it is considered as one of the suffixes in the above chart, the copula *da* behaves idiosyncratically when followed by a modal. First of all, the modal *daroo* is derived from the copula *da,* and secondly, *da* is replaced by ∅ when a modal follows (in nonpast context). This phenomenon suggests that *da* functions simply as a link or mark of relationship between one constituent and another and is less significant than other suffixes and modals in terms of syntactic relatedness and semantic representation. This issue will be addressed again in chapter 3.

In studies of modality in English, the dichotomy between proposition and modality is also discussed. For example, Austin (1962) proposed that the distinction between proposition and modality is very similar to that between the locutionary act (an act of saying something) and the illocutionary act (an act in saying something) in speech act theory. English sentence structure, however, does not exhibit a clear distinction between propositional content and modal content. The following example represents English structure:

(15) a. The teacher may cancel the class.
 b. Perhaps the teacher will cancel the class.
 c. It is all right if the teacher cancels the class.

Sentence (15)a, which is ambiguous with respect to (15)b and (15)c, illustrates that the propositional component of a sentence in English cannot necessarily be formed strictly from the constituents of the sentence. The propositional content of (15)a is, "The teacher does/will cancel the class," and the modal content is either "perhaps" or "it is all right if ~" depending on context. In Japanese, however, the propositional component is manifest in the sentence constituency, as illustrated in the following examples:

(16) Mr. Toyota should be in his office.
"It is probable that Mr. Toyota is in his office."
"Mr. Toyota is obliged to be in his office."

(17) a. Toyota san wa jimusho ni iru hazu-da.
Mr. Toyota TOP office LOC exist AUX-COP
"I am sure that Mr. Toyota is in his office."

b. Toyota san wa jimusho ni iru beki-da.
Mr. Toyota TOP office LOC exist AUX-COP
"Mr. Toyota is obliged to be in his office."

Each of the Japanese sentences contains the propositional component, *Toyota san wa jimusho ni iru*, "Mr. Toyota is in his office," with modal content expressed by either *hazu* or *beki*, each translated as "should." In English, the same auxiliary verb "should" is used to express possibility and permission, whereas in Japanese such meanings are expressed by different modals. The sentences *Toyota san wa jimusho ni iru beki-da*, "Mr. Toyota should be in his office," and *Toyota san wa jimusho ni iru hazu-da*, "Mr. Toyota should be in his office," do not convey the same meaning: the former suggests that Mr. Toyota is obligated to be in his office, while the latter expresses a speaker's assumption about Mr. Toyota's location. Thus, Japanese makes a clear distinction between the propositional and modal content, since the modal, for example *hazu* or *beki*, is located after the proposition.

One of the distinctive characteristics of modal elements in Japanese, as opposed to English, is that constituents that follow tense markers are all considered to be in the domain of modal content. This is true for final sentence particles such as the question mark *ka* and the vocative *yo*, and even the exclamation mark, since it describes how a speaker feels about the situation. The following example illustrates this relationship:

(18)

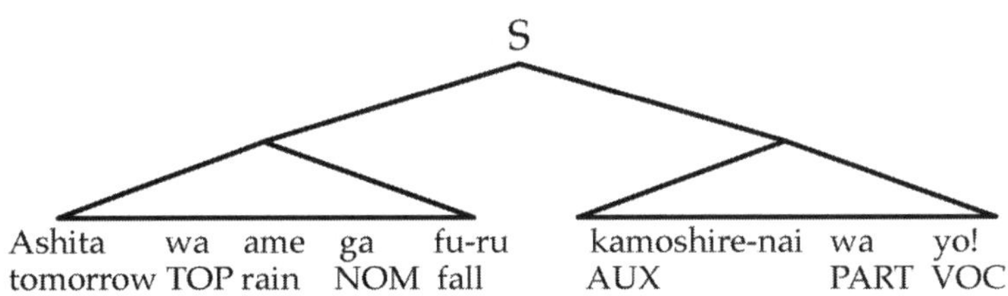

"It may rain tomorrow, I'm telling you!" (female speech)

The study of modality in Japanese is thus identical to the study of modal content—that is, it is concerned with all elements that follow the tense marker *ru* or *ta*. It is not a subject of the notions of necessity and possibility; rather, it is concerned with how a speaker views the proposition. Hence, the study of modality in Japanese is quite different from the study of modality in English.

2.3 THE CURRENT STATUS OF THE STUDY OF JAPANESE MODALITY

The study of Japanese modality received a major impetus from Tokieda's proposed distinction between propositional content and modal content. Modal content structurally involves all the constituents that follow a tense marker, even sentence final particles. Uyeno (1971) studied modality by examining these sentence final particles in terms of general semantics. She described how the use of particular sets of particles modifies an utterance in various ways to reflect the sex of the speaker, as well as the relationship, including relative social status, between the speaker and his/her addressee.

More recently, Nakau (1979) presented a clear definition of modality in terms of an opposition between proposition and modality. He defined proposition as a description of the objectified world that lies outside of the speaker, and modality as a description of the speaker's subjective attitudes which lie inside of the speaker. He also outlined some very specific features of modality, namely that: 1) modality refers only to a speaker's psychological attitude and no one else's; and 2) a speaker's psychological attitude refers only to the time of speech and not attitudes in the past or future. Based on Nakau's definition, tense and aspect are constituents of propositional content, and auxiliaries, which follow tense markers, are considered modal content. Additionally, modal auxiliaries refer to a speaker's psychological attitude at the time of speech, and can neither partake in negation nor form questions or past tense forms. These conclusions arose from the observation that the modals, h*azu,* "supposed to," and *rashii* "it seems like," for example, express modality only when they are used in the present tense. Past tense forms and past tense negative forms are, according to him, outside of the domain of modality.[15] This observa-

15. For example, when changing a sentence that includes the auxiliary *hazu* to its negative form, *hazu* itself does not partake in the negation, but instead lies in the domain of the propositional content. The same is true of *hazu* in regard to past tense. For example:

tion supports a definition of modality wherein the term refers only to a speaker's psychological attitude, and then only at the time of speech.

While Nakau focused on modals as the locus of modality, other researchers were influenced by Fillmore's theory. According to Fillmore (1968), every sentence obligatorily possesses modality, and this subsumes negation, tense, mood, and aspect. Unfortunately he did not clearly define modality, but located it in every sentence tree. Masuoka (1989a) had a view similar to that of Fillmore and claimed that modality can be found in every constituent of a sentence. Masuoka (1989b) defined modality as forms that express an agent's judgment or impression and claimed that propositional content, such as negative sentences, passive sentences, etc., can also be viewed as expressing modality since these sentences represent such speaker judgment or impression. Masuoka includes notions such as politeness, conveying thoughts, value judgment, explanation, topicalization, and so forth in addition to aspect, negation, and tense in the secondary modality category. Based on Masuoka's premise, aspect, negation, and tense can express modality, and he classified various categories depending upon the sort of psychological attitude that might be expressed in the proposition.

Moriyama (1988, 1989) divided modal content into epistemic mood and deontic mood. The influence of English linguistics begins to be seen in the use of these terms. Since Japanese sentences are recognized as consisting of propositional content and modal content, and because modal content was studied in terms of modality, it is natural that there should be some form of overlap between the two languages. Moriyama's discussion, however, does not go beyond the classification and examination of predicates and sentence types that belong to either the epistemic or deontic category.

(a)	Kare	wa	atarashii	kuruma	o	ka-u	hazu-da.		
	he	TOP	new	car	ACC	buy-NPST	AUX-COP		

"I am sure that he will buy a new car."

(b)	Kare	wa	atarashii	kuruma	o	ka-u	hazu	ga	nai.
	He	TOP	new	car	ACC	buy-NPST	AUX	NOM	NEG

"There is no way that he will buy a new car."

(c)	Kare	wa	atarashii	kuruma	o	kat-ta	hazu-da.		
	he	TOP	new	car	ACC	buy-PAST	AUX-COP		

"I am sure that he bought a new car."

As example (b) indicates, when changing *hazu* to a negative form, *hazu* itself does not form a negative predicate *(hazu-ja-nai). Hazu* and *nai* are separated by the particle *ga* and form the sentence "there is no way/possibility." The same thing can be seen in example (c): *hazu* does not take part in, but rather follows after, the past tense form of a verb.

The study of Japanese modality is thus in a state of flux, because the definition of a "speaker's psychological attitude" can involve such a wide variety of language phenomena. In Nitta's research (1989), for example, sentences such as *Kodomo ga asonde-iru*, "Children are playing," *Umi ga kirei-da*, "The ocean is beautiful," and the like are seen to include the grammatical category of modality. As such they are placed in a category with, for example, the verb *omou* "to think," and modals. Even though modality is approached from the viewpoint of propositional content and modal content, the definition makes the distinction between these two categories unclear.

Once the definition of "a speaker's psychological attitude" is applied to propositional content, even lexical items need to be examined, and we are left with the question of whether or not modality can even be expressed by words, expressions, and sentences themselves. If we approach modality from a lexical viewpoint, unlimited numbers of sentences and lexical items have to be considered to determine how each is used to express modality. For example, expressions such as *arigatoo* and *doomo*, which are the equivalent of "thank you" and "thanks," respectively, also express the speaker's attitude (they express varying degrees of politeness). The notion of an "in group" (members of one's family, company, social group and so forth) versus an "out group" (those not included in the "in group") also expresses a speaker's attitude, since it conveys a notion which is based on a given speaker's sociolinguistic background.

This type of study has most recently been undertaken by Maynard (1993) who approached modality in terms of discourse analysis and defined modality as "discourse modality."

> Discourse Modality refers to information that does not or only minimally conveys objective propositional message content. Discourse Modality conveys the speaker's subjective emotional, mental or psychological attitude toward the message content, the speech act itself or toward his or her interlocutor in discourse. Discourse Modality operates to define and to foreground certain ways of interpreting the propositional content in discourse, it directly expresses the speaking self's personal voice on the basis of which the utterance is intended to be meaningfully interpreted. (pp. 38–39)

Using this definition, Maynard examined the differences between discrete lexical items such as *dakara* and *datte* ("because"), *yahari* and *yappari* ("as expected") and *da* and *desu* ("be") and between the sentence final particles *yo* and *ne*. She claimed that language serves primarily to express subjectivity and emotion. Researchers like Maynard believe the study of modality needs to be pursued by investigating language phenomena through discrete lexical items. To these researchers this approach seems inevitable, the only way to concretely identify the borders of the concept of modality, a task which up to now has been problematic at best.

2.4 A NEW APPROACH TO JAPANESE MODALITY

As discussed in regard to the history of the study of Japanese modality, Japanese language structure is best grasped in terms of the distinction between propositional content and modal content. Although English modal content is not as clear as Japanese structurally, it is expressed by modals such as "may," "must," "should," and so forth, which are related to the notions of necessity and possibility. Japanese also has modals which are equivalent to English modals—some members of the *hoo jodooshi* "modal auxiliaries" appear in modal content. These Japanese modals have also been studied discretely by various linguists, but it was not until recently that the grammatical category of modality was established.[16] Since there is considerable common ground between English modality and Japanese modality, however, and because the English term "modality" has begun to be used frequently among Japanese researchers, it is appropriate to review what is meant by modality in English linguistics and to see whether that concept is of potential use in Japanese linguistics.

As was seen historically in the study of modality in English, modality eventually branched off from symbolic logic, as it was found that the notions of necessity and possibility could not be adequately examined from a mathematical viewpoint. Modality came to be studied in linguistics as a way of including natural language phenomena that are inevitably related to human judgment: belief, knowledge,

16. "Modal auxiliary" as a category is also found in Kitagawa and Iguchi (1988).

and a sense of morals (obligation, prohibition, and permission). Notions of necessity and possibility are then examined within the categories of epistemic and deontic modality. The terms "necessity" and "possibility" here are considered no longer as strictly alethic, but rather are based on a speaker's judgment. Modality is thus speaker-oriented and in the domain of semantics.

From this view, epistemic necessity is considered as expressing a high degree of speaker conviction that the proposition in question is true or realized. Contrarily, epistemic possibility is considered as expressing a low degree of speaker conviction toward the proposition's truth or realization. Therefore, epistemic modality is understood as expressing a speaker's assumptions or assessment of the possibilities of a proposition being true or realized. Deontic necessity, on the other hand, is considered as expressing a forceful speaker attitude in that the speaker obliges an interlocutor to realize the expressed proposition. Deontic possibility expresses that the speaker is relatively indifferent toward the proposition's realization by the interlocutor. Deontic modality, therefore, is understood as expressing an attitude toward an interlocutor that is related to the notions of permission, prohibition, and obligation. The following chart illustrates both the epistemic and deontic categories of modality. Please note that modality here is a subcategory of a larger picture of modality that is defined as "a speaker's psychological attitude":

(19)

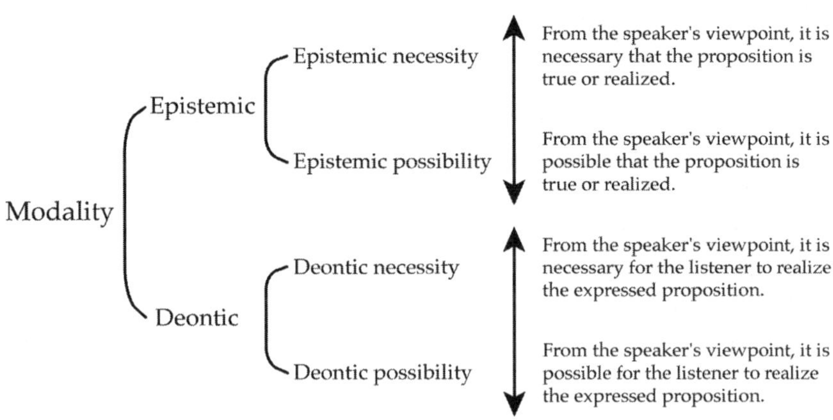

The degree of truth expressed by any of these notions is dependent on how strongly the speaker is convinced the proposition is true. The higher the degree of speaker conviction, the closer the proposition is to either epistemic or deontic necessity. The lower the degree of speaker conviction, the closer the proposition is to either epistemic or deontic possibility. For both epistemic and deontic modality, no matter how high or how low the degree of speaker conviction, the expressed proposition is always concerned with a nonactualized situation. Therefore, both notions are in the domain of possible/nonactual worlds. Modality then can be thought of as being concerned with the expression of the degree of the speaker's conviction concerning a proposition's truth or realization in the form of possible/nonactual worlds.

Filtering Japanese modals through the framework of modality taken in part from English linguistics is a plausible and practically applicable starting point for the study of Japanese language modality. In the next chapter, this exploration is begun by examining Japanese modals in the epistemic and deontic categories, which will create a framework in which it becomes possible to examine further how modality is expressed in propositional content.

CHAPTER 3

Modality and Modal Auxiliaries

3.1 JAPANESE MODALS

In chapter 2, the significance of the syntactic-semantic characteristics of propositional and modal content in Japanese was discussed, and the plausibility of applying the central notions of the study of English modality ("necessity" and "possibility") to Japanese modality was proposed. In Japanese, "mood" is expressed by modal auxiliaries (modals) that appear in modal content, distinguished from the propositional content by a tense marker *ru*, or *ta*, as a divider. Among the modals, however, not all are concerned with the notions of "necessity" and "possibility." Rather, these notions are expressed by a distinct subset of the modals, including *hazu* "must be/should be/supposed to be," *ni chigai-nai* "must be/should be," *daroo* "probably," *kamoshire-nai* "may be," *yoo(mitai)* "looks like," *soo* "appears to be/hearsay," *rashii* "seems like," and *beki* "should." These modals are therefore the primary focus of the study of Japanese modality proposed herein.

Considering the syntactic behavior of these modal auxiliaries, however, it should be noted that the commonly accepted statement that the tense markers *ru* and *ta* distinguish propositional content from modal content does not reflect their function as rigid syntactic dividers. First, the suffix *(r)u* belongs only to the lexical unit of verbs and not to adjectives and nouns, as seen in *omoshiroi* "interesting" and *sensei-da* "(s/he) is a teacher." The statement needs to be slightly modified from pertaining only to *ru* and *ta* tense markers to including imperfective and perfective markers, or simply to tense markers in general, suggesting that adjectives and nouns can also be objects of examination. Second, the modals *hazu, yoo,* and *soo* (appears to be) do not exhibit a clear divider between the proposition and the modal (e.g., *Sensei no hazu-da* "I'm sure that s/he is a teacher," *Benri na yoo-da* "It looks like it is convenient," and *Oishi-soo-da* "It looks delicious"); furthermore, the modals *ni chigai-nai, daroo,* and *kamoshire-nai* do not follow a tense marker when a sentence concerns affirmative, nonpast context (e.g., *Gakusei ni chigai-nai,* "S/he must be a student," and *Ame kamoshire-nai,* "It may be rain." Third, the modal *soo* follows a different form of lexical item in order to distinguish one meaning from the other (e.g., *Kodomo ga umare-soo-da* "The baby is about to be born" and *Kodomo ga umareru soo-da* "I hear that a baby is going to be born"). The following examples illustrate such constraints for the second and third case, for example, for sentences containing an nominal-adjectival (*shizuka* "quiet"), a nominal (*sensei* "teacher"), a verb (*furu* "to fall"), and an adjective (*omoshiroi* "interesting"):

(1) a. Asoko wa shizuka NA hazu-da/yoo-da
that place TOP quiet-COP AUX-COP
"That place is quiet, it must be/it looks like."

b. Asoko wa shizuka ∅ ni chigai-nai/daroo/kamoshire-nai/rashii/soo-da
that place TOP quiet ∅ AUX
"That place is quiet, it must be/probably/may be/seems like/I hear."

(2) a. Kare wa sensei NO hazu-da/yoo-da
he TOP teacher COP AUX-COP
"He is a teacher, must be/looks like."

b. Kare wa sensei ∅ ni chigai-nai/daroo/kamoshire-nai rashii/*soo-da
he TOP teacher ∅ AUX
"He is a teacher, it must be/probably/may be/seems like/I hear."

(3) a.

Kono	hon	wa	omoshiroI	soo-da.
this	book	TOP	interesting	AUX-COP

"I hear that this book is interesting."

b.

Kono	hon	wa	omoshiroku-nai	soo-da.
this	book	TOP	interesting-NEG	AUX-COP

"This book does not look interesting."

c.

Yuki	ga	furU	soo-da.
snow	NOM	fall	AUX-COP

"I hear that it is going to snow."

(4) a.

Kono	hon	wa	omoshiro ∅-soo-da.
this	book	TOP	interesting-AUX-COP

"This book looks interesting."

b.

Kono	hon	wa	omoshiroku-naSA-soo-da.
this	book	TOP	interesting-NEG-AUX-COP

"This book does not look interesting."

c.

Yuki	ga	furI-soo-da.
snow	NOM	fall-AUX-COP

"It looks like it is going to snow."

Some syntactic restrictions in conjoining propositional content with a modal are seen in the above sentences. They all seem to be unavoidable due to the syntactic configurations resulting from the predicate in the proposition conjoined with another constituent following it. For example, as in sentences (1)a and (2)a, the modal *hazu* originated from the noun, *"hazu,"* with the preceding constituent functioning as an attributive, modifying clause, and *na* and *no* are adjectival forms of the copula *da;* hence, the copula *da* alters the form to *na* or *no* depending on the type of constituent preceding the copula. For (1)b and (2)b, when a proposition is in past context, *dat-ta* appears before the modals, as in *Asoko wa shizuka-dat-ta ni chigai-nai* "It must have been quiet there." The absence of a copula in (1)b and (2)b is likely due to a low-level deletion rule for the nonpast tense copula in certain contexts. For sentences (3) and (4), the same morpheme *soo* is used to express two distinct meanings: "appears to be" and "hearsay." The morphological selection between the gerundive form and the nonpast tense form can be considered a device for such distinction in meaning.

Thus, although there are some syntactic irregularities observed on the border between a proposition and a modal when they are coordinated into a sentence, tense markers still maintain a role as a common syntactic divider for most modals. In addition, as seen in example (10) in chapter 2, those modals expressing a speaker's conviction are semantically independent constituents to a great extent. The role of the tense markers should still be recognized as ones that dichotomizing a proposition and a modal. This perspective permits further discussion of the concept of modality and the syntactic-semantic behavior of modals. In the following sections, the behavior of modals is examined on the basis of the categories of epistemic modality, which involve the notions of "necessity" and "possibility."

3.2 EPISTEMIC MODALITY—SUPPOSITIONALS

As stated in the previous chapter, epistemic modality is concerned with a speaker's belief and knowledge and, as such, involves notions of necessity and possibility. The behavior of each modal auxiliary in Japanese affirms that they, like their English language counterparts, are used to express notions of necessity and possibility. Specific expressions of epistemic modality include modals such as *hazu*, "must be/should be/supposed to be," *ni chigai-nai*, "must be," *daroo* "probably," and *kamoshire-nai*, "may be." These representative modals are all used to express the degree of a speaker's conviction toward a proposition, and all are provisionally termed "suppositionals" in this book. Other expressions include *yoo*, "looks like," *soo* "appears to be/hearsay," and *rashii*, "seems like," the so-called "evidentials."

As native speakers of Japanese can intuitively discern with ease, there is a degree to which these modals express notions of necessity and possibility. For instance, it is obvious to native speakers that *hazu* expresses the highest degree of necessity, that *kamoshire-nai* expresses the lowest degree of possibility, and that the rest of the auxiliaries lie somewhere in between. Evidentials, however, have not yet been examined from the viewpoint of these two notions. In order to investigate the degree to which suppositionals and evidentials express necessity and possibility, first, a brief examination of each discrete modal is given.[1] Then, both suppositionals and evidentials are examined based on: i) compatibility with adverbs that are used to express degree of speaker conviction; ii) compatibility with the actual situation described in the second half of the sentence; and iii) compatibility with another modal.[2] These examinations are based on a survey that was conducted using a total of forty Japanese people, consisting of twenty males and twenty females, between the age of thirty and forty, living in Tokyo, who do not speak English. Also, please note that in this descriptive examination of modals, only reportive-style narratives were the object of the examination, and nonreportive style narratives were not integrated into the semantic analysis.

3.2.1 *Hazu* (must be, supposed to be)

Hazu[3] can function as either a suppositional or a nominal,[4] exhibiting such meanings as "must be," "ought to be," "expected to be," or "possibility," "reason," "plan." In other words, suppositional *hazu* is a modal, whereas nominal *hazu* is not. Morita (1980) asserted that *hazu* is used when a speaker makes a judgment based on objective grounds. The suppositional *hazu* lies outside of a proposition, and it indicates to the interlocutor that the speaker had firm evidence about the matter in question before s/he uttered the proposition. It expresses a high degree of speaker conviction, as illustrated in the following examples:

1. The discussion of each modal is also found in Teramura (1979, 1984), Nakau (1979), Morita (1980), and Makino & Tsutsui (1989).
2. For example:

(a)

Taroo	wa	sono hi	hayaku	uchi	e	kaet-ta.	
Taro	TOP	that day	early	home	LOC	return-PAST	

"Taro returned home early that day."

Hachiji	ni	Hanako	kara	denwa	ga	kur-u	hazu-dat-ta.
8 o'clock	TEMP	Hanako	from	phone call	NOM	come	AUX-COP-PST

"He was supposed to receive a phone call at 8 o'clock from Hanako."

(b)

Hachiji	ni	Hanako	kara	denwa	ga	kuru	hazu-da.
8 o'clock	TEMP	Hanako	from	phone call	NOM	come	AUX-COP

"There should be a phone call from Hanako at 8 o'clock."

Hachiji ni Hanako kara denwa ga kuru hazu-dat-ta "He was supposed to receive a phone call from Taro at 8 o'clock" in sentence (a) represents Taro's internal feeling (expressed in sentence [b]) and takes the past tense form due to the fact that the entire narrative is in the past tense form.

3. Detailed discussion of *hazu* is most recently found in Kinoshita (1997).
4. When *hazu* is used as a noun, it is in the domain of propositional content as is represented by the following sentence:

(5) Saifu wa tsukue no kihidashi ni aru hazu-da.
wallet TOP desk GEN drawer LOC exist AUX-COP
"The wallet must be in the desk drawer."

Sentence (5) is taken from a conversation where the speaker previously confirmed the existence of the wallet in the desk and, therefore, takes the fact that the wallet is still there for granted. If the speaker simply seeks to convey the wallet's location to the interlocutor, a simple statement could be used without adding *hazu: Saifu wa tsukue no hikidashi ni aru* "The wallet is in the desk." In sentence (5), however, this is not the case. *Hazu* is used, presumably, because the interlocutor at first could not find the wallet in a place where it was supposed to be. The following is the complete version of an actual conversation:

(6) A: Nee, saifu doko?
Say, wallet where
"Say, where is the wallet?"

B: Tsukue no hikidashi no naka yo.
desk GEN drawer GEN inside VOC
"It is in the desk drawer."

A: E, sagashi-ta kedo, nakat-ta yo.
huh look for-PST but NEG-PST VOC
"Huh? I looked for it, but I could not find it."

B: Sonna koto nai wa yo.
such a thing exist-NEG PART VOC
"That cannot be.

Sakki katazuke-ta toki shimat-ta kara,
little while ago clean-PST when put away-PST and so
"When I cleaned I put it away, so . . ."

Kare ga sonna koto o iu hazu wa nai.
he NOM such thing ACC say reason CNT NEG
"There is no reason (way) for him to say things like that."

In this sentence, *hazu* functions as a subject, with *kare ga sonna kot o iu* "He says things like that" as an appositional clause modifying it; therefore, *hazu* can be replaced by another noun such as *wake* or *doori,* both of which mean "reason." This *hazu* has nothing to do with speaker supposition, but merely reflects a factual statement.

hikidashi ni aru hazu yo.
drawer LOC exist AUX VOC
"I'm sure that the wallet is in the (desk) drawer."

A: Hen-da na. Ja moo ichido sagashite-miru.
strange-COP PART well one more time look for-try.
"That's strange. Well, I'll take another look."
A, at-ta, at-ta.
Ah, find-PST find-PST
"Ah, I found it. I found it."

In the above conversation, speaker B first stated the location of the wallet using a simple statement "It is in the desk." This use of language shows that speaker B is confident in his/her knowledge of the wallet's location. When this knowledge was challenged by speaker A, however, speaker B used *hazu* in his/her second reply. This is due to the information gap presented by speaker A, which reduced speaker B's confidence in the proposition. Although speaker B could have repeated the same statement, the speaker's slight hesitation caused him/her to avoid using the factual statement. Speaker B then tried to convince the interlocutor based on his/her knowledge by presenting an objectively observable event (in this case, the fact that speaker B had put the wallet in the desk drawer) that supports speaker B's statement to be true. Thus, when *hazu* is used, there is usually either presentable evidence or a speaker's knowledge that supports the speaker's judgment. Such preacquired evidence/knowledge creates a spatiotemporal/psychological distance between the speaker and the evidence/knowledge and allows the speaker to perceive the evidence externally and objectively. This construction is a way for a speaker to express his/her high conviction that the proposition is true.

Depending on a speaker's perception, presentable evidence can be good enough for the speaker to utter a factual statement, that is, it is not necessary to add *hazu* to the proposition. In this sense, *hazu* is, at root, associated with the concept of "plan," and can often be restated by *-kotoni-natte-iru* "it has been decided," the latter of which is not a modal, but simply constitutes a proposition. The difference between the two is that while *hazu* always reflects a situation where the speaker forms a conjecture based on firm evidence, *-kotoni-natte-iru* is used when a statement is concerned with a fact that is not subject to conjecture or interpretation on the part of the speaker. The key to using either of these forms lies in context as well as the degree of the speaker's belief toward thc proposition.[5] The following examples clarify the difference between *hazu* and *-kotoni-natte-iru:*

(7) A: Miwa san saikin isogashi-soo ne.
Miwa recently busy-AUX PART
"Miwa looks busy recently."

5. Kuno (personal communication 2001) suggests that *-kotoni-natte-iru* must involve man-made principles/rules/plans, etc. The following examples illustrate the difference between *hazu* and *-kotoni-natte-iru:*

A (a) Raigetsu kara kyuuryoo ga deru hazu-da.
Next month from salary NOM be paid AUX-COP
"I'm sure that I will get paid starting next month."

(b) Raigetsu kara kyuuryoo ga deru kotoni-natte-iru.
Next month from salary NOM be paid-become-PERF
"It has been decided that I will get paid starting next month."

B (a) Kongetsu no shuunyuu kara
this month GEN income from
shishutsu o hiku-to zero ni-naru hazu-da.
expense ACC subtract-COND zero-become AUX
"I'm sure that it becomes zero when the expense is subtracted from the income of this month."

* (b) Kongetsu no shuunyuu kara shishutsu o
this month GEN income from expense ACC
hiku-to zero ni-naru-koto ni-natte-iru.
subtract-COND zero-become-COMP-become-PERF
"It has been decided that it becomes zero when the expense is subtracted from the income of this month."

C (a) Yamada sensei wa raigetsu 70 sai ni-naru hazu-da.
Yamada teacher TOP next month 70 years old-become AUX-COP
"I'm sure that Professor Yamada will become 70 years old next month."

* (b) Yamada sensei wa raigetsu
Yamada teacher TOP next month
sai ni-narukoto ni-natte-iru.
years old-become-become-PERF
"It has been decided that Professor Yamada will become 70 years old next month."

While both *-kotoni-natte-iru* and *hazu* are appropriate for sentences A(a) and A(b) where the event can be controlled by a human force, *-kotoni-natte-iru* is inappropriate when an event occurs inevitably or occurs as a natural course of events. The modal *-kotoni-natte-iru* is an equivalent of "it has been decided" and implies that the decision was made by some other source that has willful power to bring a settlement. These examples support Kuno's suggestion.

B:	Raigetsu	kekkon surukotoni-natte-iru-n-da.
	next month	expected to get married-become-PERF-COMP-COP

"She is expected to get married next month."

(8)	A:	Miwa san	rokugatsu	ni	kekkon suru	soo	ne.
		Miwa	June	TEMP	get married	AUX	PART

"I hear that Miwa is going to get married in June."

B:	E,	iya,	kugatsu	no	hazu-da	kedo . . .
	really	no	September	GEN	AUX-COP	but . . .

"Really? I 'm sure that she is getting married in September . . ."

In sentence (7), speaker B is relaying second-hand information that does not necessarily involve the speaker's inference or judgment about whether or not Miwa is getting married. The situation in (8), on the other hand, is one that causes speaker B to question what s/he knows. There is an inconsistency between speaker B's knowledge and speaker A's claim, and this gap affects the level of confidence that speaker B has in the proposition. Thus, *hazu* can be used instead of *-kotoni-natte-iru* in a situation where the confidence a speaker has in the validity of his/her proposition deteriorates to the point that the speaker him/herself no longer considers his/her own information to be knowledge, but rather in his /her own mind that knowledge is reduced to the status of belief.[6] *Hazu* vacillates between statement and judgment due to a perceived inconsistency. Consequently *hazu,* in present context, is equivalent to the English phrase "I am sure that . . ." or "in light of what I know it is necessarily true that . . ."

Hazu can also appear in negative or past tense forms, in which cases it no longer functions as a suppositional. As Teramura (1979) and Nakau (1979) both suggest, modality refers to a speaker's psychological attitude at the time of speech and it cannot exist in negated,

6. For this reason, *hazu* is often inappropriate when the speaker talks about his/her own plan as in the following example:

Watashi	wa	ashita	Nihon	e	iku	hazu-desu.
I	TOP	tomorrow	Japan	LOC	go	AUX-COP Q

"I'm sure that I'm going to Japan tomorrow."

This sentence suggests a speaker's unawareness of his/her own activity.

questioned, or past tense forms. Therefore, *hazu,* in such an environment, communicates the concept of *-kotoni-natte-i-ta,* "it was decided," which is clearly a proposition. The following examples represent the use of *hazu* in past and past negative contexts:

(9)	Miwa	wa	kinoo	Tookyoo	e	kaeru	hazu-dat-ta.
	Miwa	TOP	yesterday	Tokyo	LOC	return	AUX-COP-PST

"Miwa was expected to return to Tokyo yesterday."

(10)	Fuyuhiko	wa	Sendai	e	kuru	hazu-ja-nakat-ta.
	Fuyuhiko	TOP	Sendai	LOC	come	AUX-COP-NEG-PST

"Fuyuhiko was not expected to come to Sendai."[7]

Sentences (9) and (10) convey counterfactual situations, expressing that something was once decided or preplanned, but that the plan was not realized, and *hazu* is not used as a suppositional. In this case, the past tense forms of hazu are semantically identical to *-kotoni-natte-i-ta* "he was supposed to" and *-kotoni-natte-i-nakat-ta* "he was not supposed to," respectively. The meaning is no longer the same as that of the suppositional *hazu;* it no longer expresses a speaker's strong belief, but expresses an expectation. When *hazu* is used as a suppositional, the situation is one where a speaker tries to express his/her belief based on firm evidence. Other uses of *hazu* indicate a speaker's expectation, or draw attention to some planned event.

Hazu expresses epistemic modality only when it appears in modal content and relates to epistemic necessity rather than epistemic possibility. When *hazu* is viewed in terms of the notions of necessity and possibility, the notion of possibility alone does not fully reflect what *hazu* is meant to express. This is because possibility does not have to be based on firm evidence, but can simply convey a speaker's intuitive supposition toward a proposition. This, however, is not what *hazu* is meant to express. Necessity, on the other hand, demands that a proposition be true based on what the speaker knows. *Hazu* is not belief without evidence; rather, the belief has to be derived from

7. This sentence has the same meaning as *Ko-nai hazu-dat-ta.* When *hazu* is part of a past tense construction, it creates a counterfactual meaning and can also partake in the negative form. This is true especially when *hazu* is in a modifying clause. For example, *Kuru hazu ja-nakat-ta hito ga ki-ta,* The person who was not supposed to come, came."

something a speaker witnesses or hears, and then processes into some sort of personal knowledge, and it always requires evidence to support its claim. The evidence can be based upon the speaker's observations or upon something which has previously been accepted as fact. Therefore a sentence with *hazu* may often be restated, *-kotoni-natte-iru* "something has been decided or something is supposed to" if a speaker is in a situation where s/he is to give a factual statement. The following chart illustrates the function of *hazu* in various contexts:

(11)

		hazu				
		Non-Past	Non-Past Neg.	Past	Past Neg.	
iku	Non-Past	*Iku hazu da.* I am sure that he is going.	*Iku hazu ja-nai.* He is not supposed to go.	*Iku hazu dat-ta.* He was supposed to go.	*Iku hazu ja-nakat-ta.* He was not supposed to go.	Non-Modal
	Non-Past Neg.	*Ik-anai hazu da.* I am sure that he is not going.	*Ik-anai hazu ja-nai.* He is not supposed to not go.	*Ik-anai hazu dat-ta.* He was supposed to not go.	*Ik-anai hazu ja-nakat-ta.* He was not supposed to not go.	
	Past	*It-ta hazu da.* I am sure that he went.				
	Past Neg.	*Ik-anakat-ta hazu da.* I am sure that he did not go.				
		Modal				

It is manifest from the table that *hazu* appearing in a nonpast context is necessarily a modal (light grey regions), whereas *hazu* in any other context is nonmodal (dark grey regions). Furthermore, whereas a verb occurring in nonpast context (affirmative or negative) can accompany either modal or nonmodal *hazu,* a verb in past context occurs only with the modal version. Finally, note the inappropriateness of the following sentences, corresponding to the "X" region of the table:

*(a) It-ta hazu-ja-nai.
"He is not supposed to have gone."

*(b) Ik-anakat-ta hazu-ja-nai.
"He is not supposed to have not gone."

*(c) It-ta hazu dat-ta.
"He was supposed to have gone."

*(d) Ik-anakat-ta hazu-dat-ta.
"He was supposed to have not gone."

*(e) It-ta hazu-ja-nakat-ta.
"He was not supposed to have gone."

*(f) Ik-anakat-ta hazu-ja-nakat-ta.
"He was not supposed to have not gone."

However, the structure in sentences (a) and (b) altered to, for example, *it-ta hazu wa nai* "There is no way that he went," is appropriate, where *hazu* is used as a nominal and is not a modal.

3.2.2 *Ni chigai-nai* (must be)

Ni chigai-nai[8] is used when a speaker is convinced that a proposition is true. According to Morita (1980, p. 412), *ni chigai-nai* is used under the same notion as the phrase *machigai wa nai* "there is no mistake" and is used to express a speaker's firm belief that the proposition is true or realized. In this sense, *ni chigai-nai* and *hazu* are very similar in their use, and both are often interpreted as the English equivalent of the modal "must." The main difference between *ni chigai-nai* and *hazu*, however, is that *ni chigai-nai* only expresses a speaker's strong sense-based, intuitive observations about asserting the truth of a proposition, while *hazu* involves a perceptual distance between the speaker and the acquired evidence that is objectively perceptible. This observation suggests that *hazu* is closer to the notion of epistemic necessity than *ni chigai-nai*. The following examples demonstrate the differences between these two suppositionals:

(12)	Kyoo	wa	depaato	de	seeru	ga	aru	kara
	today	TOP	department store	LOC	sale	NOM	exist	and so

8. English gloss "AUX" is given for *ni chigai-nai* and *kamoshire-nai* even though both contain the negative morpheme *nai* "NEG." They are both considered modals in terms of the English gloss.

haha	wa	imagoro	kaimono	o	shite-iru	ni chigai-nai /hazu-da.
mother	TOP	now	shopping	ACC	do-PROG	AUX/AUX-COP

"Since there is a sale at the department store today, my mother must be shopping (there) now."

(13)

Tabako	no	suigara	ga	aru	kara,
Cigarette	GEN	stub	NOM	exist	because

dareka	ga	i-ta	ni chigai-nai/hazu-da.
someone	NOM	exist-PST	AUX/AUX-COP

"Since there are cigarette butts, someone must have been (here)."

(14)

Atama	ga	itai	shi,	netsu	mo	aru	tokoro	o	miru-to
head	NOM	hurt	and,	fever	too	have	state	ACC	observe-COND

kaze	o	hii-ta	ni chigai-nai/??hazu-da.
cold	ACC	catch-PST	AUX

"Since I have a headache and fever, I must have caught a cold."

(15)

Hadaka de	asonde-iru	kara	kaze	o	hiku	ni chigai-nai/ ??hazu-da.
naked	play-PROG	since	cold	ACC	catch	AUX/AUX-COP

"Since (the child) is playing naked, s/he must catch a cold."

(16)

Sooiu koto	itte-iru	kara,	atode	okor-areru	ni chigai-nai/ ??hazu-da.
such thing	say-PROG	since	later	scold-PASS	AUX/AUX-COP

"Since he is saying things like that, I'm sure that he will be scolded later."

In sentences (12) and (13), there is evidence that is processed and observable as the result of an action (e.g., the acquired knowledge of the sale at the department store and the presence of cigarette butts, respectively), and either suppositional is appropriate regardless of attenuate differences detected therein.[9] *Hazu,* however, is an unacceptable (or at least awkward) option for sentences (14)–(16), due in part to the fact that the basis for judgment occurs in the presence of the speaker. As discussed in section 3.2.1, *hazu* creates a spatiotemporal distance between the speaker and the situation described in the proposition, allowing him or her to perceive the situation objectively at the time of speech. In order for the speaker to view the situation externally, however, the proposition must be a logical consequent of the antecedent, and the consequent cannot chronologically precede the antecedent. When the speaker is in the midst of an ongoing situation, such chronological order is yet to be established; therefore, sentence (14) is unacceptable.[10] Sentences (15) and (16) relate to the present moment and/or point toward a future for which the speaker has not the ability to observe the situation externally. Thus, *hazu* is not appropriate when an objective decision cannot yet be made regarding the outcome—that is, the condition has not yet become objectively perceptible as a basis for judgment. When the basis for judgment exists solely in the present situation, *ni chigai-nai* is used.

It can be concluded that *ni chigai-nai* is used to express the speaker's conviction (based on outward evidence or the speaker's speculative intuition) that a proposition has a strong possibility of

9. Teramura (1984) suggests that the difference between these types of situations is that *ni chigai-nai* is often used in a monologue that does not invite an interlocutor in a conversation, while *hazu* tends to be used when an interlocutor is present. This is likely due to the fact that the interlocutor may plant some doubt in the speaker's mind about the truthfulness of the propositional content of the *hazu* sentence.

10. When the speaker is in the midst of the ongoing situation, s/he cannot view the whole event externally and *hazu* is not an appropriate choice. The use of *hazu* becomes less awkward when the consequence of an ongoing situation is considered a previously established principle, since such allows the speaker to predict the consequence without actually viewing the situation objectively. The following is one such example taken from commonly acknowledged lore regarding the weather:

Nishi	no	sora	ga	danndann	akaku-natte-ki-ta	kara
west	GEN	sky	NOM	gradually	red-become-come-PAST	since
ashita	wa	harer	hazu-da.			
tomorrow	TOP	sunny	AUX-COP			

"Since the sky in the west is gradually becoming red, I'm sure that it will be sunny tomorrow."

In the above sentence, the speaker witnesses the change in the color of the sky, and the speaker's preacquired knowledge regarding weather patterns is the basis for such judgment.

being realized. *Hazu*, on the other hand, requires hard evidence (rather than intuition) and therefore expresses a higher degree of epistemic necessity.

Ni chigai-nai can partake in the past tense form, but unlike *hazu*, it does not invite a strong counterfactual interpretation. *Ni chigai-nakat-ta* indicates a speaker's firm belief that the proposition was true at some time in the past, but it does not necessarily mean that the speaker continues to believe it as firmly. The following examples compare such use of *ni chigai-nai* with *hazu:*

(17) a. Sore wa daiamondo no hazu-dat-ta.
that TOP diamond GEN AUX-COP-PST
"It was supposed to be a diamond."

b. Sore wa daiamondo ni chigai-nakat-ta.
that TOP diamond AUX-PST
"I was sure that it was a diamond."

Sentence (17)a clearly conveys that the article was not a diamond, but something else—a cubic zirconium, for example. *Ni chigai-nakat-ta*, on the other hand, expresses that the speaker's conviction is drawn out over an extended period up to the time of speech and that his/her conviction was certain at the time of judgment. It could be the case that the speaker believes that the article was a diamond at the time of speech, or that the speaker believed the article to be a diamond, but it does not offer a clear picture of the actual situation.[11] From sentence (17)b alone, the interlocutor would not know clearly whether or not the article is a real diamond. *Ni chigai-nakat-ta* thus does not quite objectify the past event or the speaker's judgment of that event

11. Note the following examples:

(a) Kinoo wa arashi ga kiihantoo e kuru hazu-dat-ta.
yesterday TOP storm NOM Kii peninsula LOC come AUX-COP-PST
"The storm was supposed to come to the Kii peninsula yesterday."

(b)? Kinoo wa arashi ga kiihantoo e kuru ni chigai-nakat-ta.
Yesterday TOP storm NOM Kii peninsula LOC come AUX-PST
"I believed that the typhoon would come to the Kii peninsula yesterday."

to the same extent as *hazu*. This observation also suggests that *ni chigai-nai* may function as a modal in past tense form.

Although the function of *ni chigai-nai* is somewhat similar to that of *hazu*, the former can also be used to express a speaker's intuitive judgment which does not have to be based on firm evidence. The speaker can simply express a strong belief based on intuition as well as presentable evidence. For example, a speaker may use *ni chigai-nai* rather than *hazu* for topics such as ghosts, UFOs, and so forth for which we do not have firm, scientific evidence. One might say *UFO wa iru ni chigai-nai* "UFOs must exist," but *UFO wa iru hazu-da* may be rather awkward unless the speaker objectively witnesses a UFO or scientifically proves the existence of UFOs. Consequently, *ni chigai-nai* is closer to the notion of possibility than that of *hazu*. *Ni chigai-nai* can be used in a much wider variety of situations where a speaker confidently infers that a proposition is true.

3.2.3 *Daroo* (probably is)

Daroo (and its polite equivalent *deshoo*)[12] is typically used when a speaker wishes to avoid the tone of directness associated with the use of *da*, creating a sentence that conveys a somewhat softer, more polite impression. *Daroo* began to attract increasing attention among linguists as the study of modality became more widespread in Japanese language research. Moriyama (1989), for example, posited that *daroo* exhibits two meanings: confirmation of a proposition (the same function carried by the tag part of a tag question in English; this meaning is similar to that of the sentence final particle *ne*, as discussed later in this section) and judgment of probability following a proposition. Which meaning is appropriate depends upon whether or not the interlocutor possesses the information in question: when the interlocutor possesses the same information as the speaker, *daroo* is used to confirm that the interlocutor agrees with the information; when the

---

Sentence (a) conveys that the storm did not in fact hit the *Kii* peninsula. Since the outcome of the past event can be judged from the present, it is easy to view that actual situation from this sentence. Sentence (b) is, however, somewhat awkward as it gives the impression that the speaker still does not know the actual situation. In daily conversation, such uncertainty creates awkwardness.

12. *Daroo*, an informal level expression, is a contraction of the formal suppositional form of *da*, and *deshoo* is the suppositional form of *desu*, the polite equivalent of *da*.

speaker presupposes that the interlocutor does not have the information, *daroo* is used to express the speaker's judgment with respect to the proposition. In the former case, the function of *daroo* is identical to that of the sentence final particle *ne.* The key in distinguishing between the two meanings is usually by way of intonation, as illustrated in the following examples:

(18) a. Dame-da-tte it-ta daroo/deshoo?
no-COP-QUOT say-PST COP/COP(POL)
"I told you 'no,' didn't I?"

b. Dame-da-tte it-ta ne?
no-COP-QUOT say-PST PART
"I told you 'no,' didn't I?"

(19) Ashita wa tabun ame daroo/deshoo.
tomorrow TOP probably rain AUX/AUX(POL)
"It will probably rain tomorrow."

Sentences (18) and (19), if no context or intonation is provided, are each ambiguous between the "tag question" and "judgment of probability" interpretations. It is the interrogative intonation in (18) that invokes the former meaning, and the lack thereof in (19) that invokes the latter.

The use of *daroo* can further be investigated by applying Kamio's (1990, 1994, 1997a, 1997b) theory of territory of information, which argues that language use depends on the territory to which information described by a sentence belongs. In this theory, whether or not a speaker/interlocutor possesses a given piece of information (i.e., whether s/he knows it or not) is clearly distinguished from whether or not a piece of information belongs to his/her territory. Kamio used the terms "direct form" and "nondirect form" for such a distinction. He hypothesized that "the direct form is used when the speaker has adequate evidence for an assertion" and "is appropriate when the information the form expresses is deep within the speaker's territory of the information," whereas "the nondirect form is used when the speaker's evidence is insufficient" and "is appropriate when the information falls less deeply within the speaker's territory or even outside it" (1997, p. 146).[13] The territory to which a given piece of information belongs depends on the degree of the speaker's and the interlocutor's familiarity and involvement in the situation expressed by the sentence. For example, if the speaker lives in Tokyo, and the interlocutor does not, the speaker can legitimately assume that the information that Tokyo's population is 12,000,000, belongs to his/her territory. Therefore, s/he can legitimately use the bare-form sentence *Tookyoo no jinkoo wa 12,000,000* (sen-nihyakuman) *kurai desu,* "The population of Tokyo is about 12,000,000" to represent this information. On the other hand, if the interlocutor lives in Tokyo, and the speaker lives in a neighboring town, s/he cannot legitimately claim that the information belongs to his/her territory because the interlocutor's familiarity with, and involvement in, this piece of information are stronger than the speaker's. (The fact that the speaker knows the truthfulness of this proposition is not sufficient for him/her to claim that the information belongs to his/her territory, as well as to the interlocutor's.) Therefore, in such a situation, the speaker cannot use the bare-form sentence to convey the information to the interlocutor, but must use sentences ending with forms that signal that the information does not belong to his/her territory. *Rashii* and *yoo* are two such examples.

Kamio (1990) applied the theory to the use of *ne* (tag question marker), which has the same function as *daroo* as exemplified in sentence (18)b. His investigation concluded that *ne* is an indicator that information belonging to the speaker's territory is identical to that belonging to the interlocutor's territory at the time of speech.[14] This is why *ne* is not used when a speaker is asking for an answer in a

13. Kamio used the following examples to distinguish the "direct " and "indirect" forms.

A. (a) The man is a police detective.
(b) It looks like the man is a police detective.

B. (a) This book sells extremely well.
(b) I hear that this book sells extremely well.

A(a) and B(a) are both "direct forms," while A(b) and B(b) are "nondirect forms." He also included forms such as "Is that so?" and "If you come, we can go to a coffee shop" in the category of nondirect forms.

14. In his later discussion (1994, p. 96), this *ne* is referred to as "obligatory *ne*," as opposed to "optional *ne*." The principles of obligatory *ne* stipulate that "the speaker must use *ne* when s/he assumes that a given piece of information falls equally, or more deeply (even exclusively), within the hearer's territory, and to the fullest degree.

simple question sentence, such as "How much is it?" (*Ikura-desu ka,* not *ikura-desu ne* in Japanese). Thus, when *daroo* is used as a tag question marker, it is not used to express the degree of speaker supposition toward the expressed proposition; rather, it confirms somewhat rhetorically the truth of the proposition through the interlocutor. Therefore, the function of this *daroo* is distinct from the function of suppositional *daroo.*

Daroo, however, has a characteristic different from the suppositionals *hazu,* and *ni chigai-nai* (and *kamoshire-nai,* discussed in the next section), in that it can invite either *ka* or *ne* to form a question sentence. The following example illustrates this feature:

(20) a. Doo sure-ba ii ka
how do-COND good Q
"What shall I do?"

* b. Doo sure-ba ii ne.
how do-COND good PART
"What shall I do, right?"

(21) a. Doo sure-ba ii-daroo ka.
how do-COND good-AUX Q
"What do you think I should do?/I wonder what I should do."

b. Doo sure-ba ii-deshoo ka.
how do-COND good-COP Q
"What do you think I should do?"

c. Doo sure-ba ii-daroo/deshoo ne.
how do-COND good-AUX/COP Q
"What shall I do (I wonder)?"

The proposition *Doo sure-ba ii,* "What shall I do," is a question sentence that clearly indicates that the speaker does not have the desired information/answer; therefore, *ne* is inappropriate as seen in sentence

(20)b. On the other hand, *daroo* can appear with either *ne* or *ka* for the same proposition, as seen in (21)a and (21)c.

As the English equivalents show, sentence (21)a can be interpreted either as a question addressed to the interlocutor, or as a monologue (i.e., a question addressed to the speaker him/herself). When *daroo* in sentence (21)a is replaced by its polite equivalent *deshoo,* as in sentence (21)b, only the "question addressed to the interlocutor" interpretation is granted. This means that in a question sentence, *deshoo* is used only in association with an interlocutor, whereas *daroo ka* may be used when the speaker and interlocutor are identical. This use of *daroo ka* creates the meaning "I ask myself (I wonder)" when the question is uttered toward the speaker him/herself, and an interlocutor is not necessarily present. The reason that *deshoo ka* has only one interpretation is similarly explained: there is no reason for the speaker to be overly polite upon asking him/herself a question. Thus, *deshoo ka* necessarily involves an interlocutor and functions as the polite equivalent of *desu ka.*

Daroo in sentence (21)c, however, accompanies *ne,* and according to Kamio's description, this *ne* does not function as a tag question marker, since the proposition includes a question word (*doo*) which indicates that the speaker does not possess an answer. Inoue (1990) examined the function of *daroo ne* and concluded that it functions the same as *desu ka.* The examples of sentences (21)a–(21)c not only support his idea, but also lead one to the conclusion that both *daroo ka* in sentence 21(a) and *daroo ne* in sentence 21(c) are functionally equivalent to *desu ka* and, hence, such use of *daroo* does not function as a suppositional.

Based on the theory of territory of information, suppositional *daroo* belongs to territory outside the speaker's domain due to uncertainty in the truth of information. Suppositional *daroo* often conveys a relatively low level of speaker confidence as to the correctness of the proposition under consideration. The following sentences demonstrate this aspect of *daroo:*

(22)	Okane	ga	nakute-wa	shinsha	wa	ka-e-nai	daroo.
	money	NOM	not-if	new car	CNT	buy-POT-NEG	AUX

"If you do not have money you probably cannot buy a new car."

(23) Anohito wa watashi no koto o moo wasurete-iru daroo.
he TOP I GEN things ACC already forget-PERF AUX
"He probably has already forgotten me."

Daroo in the above sentences simply expresses a speaker's conjecture based on the environment in which the speaker delivers judgment, which can be based on firm evidence or a speaker's intuition. For example, in sentence (22), based on the fact that someone has no money, the speaker surmises that a new car probably cannot be purchased. In this sentence, *daroo* implies strong support for the possibility of that proposition being true. Contrarily, sentence (23) can be spoken in various contexts, as no basis for the judgment is given. *Daroo* may simply express the speaker's intuition, or the speaker may have some other support for making the statement.[15] This premise leads to the conclusion that the degree of speaker conviction is not always clearly specified by the presence of *daroo.* Not surprisingly *daroo* sentences are often taken to be noncommittal statements.

The function of *daroo* is quite complicated, and due to its syntactic features, it is often context-dependent. However, when *daroo* is used as a suppositional, it indicates a relatively wide range of implications, involving both necessity and possibility, in regard to the speaker's confidence in the truth of the proposition. Accompanying adverbs, such as *tabun* "probably," *osoraku* "perhaps," and *moshikashi-tara* "maybe" can generally clarify the meaning of *daroo* in these cases. This issue is taken up in section 3.4.1 with regard to compatibility tests between modals and adverbs.

3.2.4 *Kamoshire-nai* (may be)

Kamoshire-nai literally means "it cannot be known." Its root form is fundamentally negative and it acts in much the same way as an adjective. Morita (1980, p. 224) states that *kamoshire-nai* is used when a

15. The argument that *daroo* is unequivocally speaker-centered is also given in section 3.4.3. In that section, the compatibility of modals with another modal is tested, and hearsay *soo* is shown appropriate in following another modal in order to express a third party's conviction/judgment. *Daroo*, however, is the only exception, obstinately defending its function of expressing only the speaker's supposition (see footnote 33).

speaker is uncertain about the truth of a proposition and therefore presents his/her own hypothetical idea. He states that the probability of the proposition being true is low since it is based solely on the speaker's intuitive judgment.

Kamoshire-nai can be attached to the past and nonpast tense forms of verbs and adjectives, and it selects a zero-form copula when attached to nominal and nominal-adjectival predicates. *Kamoshire-nai* cannot invite a question mark, but it can be used in the past tense form. Its past tense forms, however, may present features different from the past tense form of *hazu* and *ni chigai-nai.* Let us first examine *kamoshire-nai* in present context:

(24)	Rainen	atari	ookii	jishin	ga	kuru	kamoshire-nai.
	next year	around	big	earthquake	NOM	come	AUX

"A big earthquake may come sometime next year."

(25)	Kare	wa	kyoo	no	kaigi	ni	shussekishi-nai	kamoshire-nai.
	he	TOP	today	NOM	meeting	DAT	attend-NEG	AUX

"He may not attend today's meeting."

Kamoshire-nai can be used when a speaker does not possess firm evidence, but judges a situation based on his/her intuition and the surrounding environment. Therefore, *kamoshire-nai* is often used in reference to totally unknowable, imagined situations that are in the domain of epistemic possibility. A sentence with *kamoshire-nai* can be interpreted as "it is possible that X happens," and it is simultaneously compatible with "it is possible that X does not happen." *Kamoshire-nai* neither implies that X will happen nor implies that X will not happen, but is compatible with both. For the most part it is used to express situations whose outcome cannot be determined by the speaker, that is, indeterminable situations. Therefore, *kamoshire-nai* reflects the hypotheticality of the proposition.

The term "hypotheticality" usually refers to a logical hypothesis—a tentative assumption made in order to draw out and test its logical or empirical consequences. This is represented by conditional sentences, which do not have truth value, that is, the constituent propositions are not known to be true. They are neither determinable nor controllable and can relate to unknown, unrealized future situations. This can be applied in empirical conjecture, especially in unknown, unrealized future situations. In this sense the suppositional *kamoshire-nai,* which is used to express that the speaker has no basis for the proposition's truth, can be said to express the notion that the proposition has neither truth value, nor expresses determinability; therefore, it expresses hypotheticality.

Hypothetical situations often cause a speaker to imply or an interlocutor to infer that a result other than that expressed in the antecedent is possible due to the unpredictability of the realization of the expressed antecedent. For example, in the use of natural language, if one says, "If you give me 10 dollars an hour, I would work for you," it invites an inference of the opposite situation—"If you do not give me 10 dollars an hour, I would not work for you." This tendency is referred to as an "invited inference" proposed by Geis and Zwicky (1971), who pointed out the applicability of biconditionals to natural language phenomena and suggested that a sentence of the form $X \supset Y$ invites an inference of the form $\sim X \supset \sim Y$. Although they concluded that there is no evidence of a direct relationship between invited inferences and syntactic form, it is not difficult to perceive that the higher the degree of hypotheticality of a sentence, the stronger the possibility of the proposition's suggesting invited inference.[16] In other words, when a speaker has firm evidence upon stating a proposition, the statement would not cause an interlocutor to infer an alternate situation. On the other hand, if the speaker is uncertain of the situation being true or realized, the interlocutor could not help inferring an alternate situation, as seen in the above example. This is related to epistemic possibility, and refers to a possible/nonactual world where an alternate situation can easily be insinuated.

Interpretation of *kamoshire-nai* as hypothetical is also valid when *kamoshire-nai* partakes in the past tense form. Although the outcome of the proposition in past context is obvious, since the past event is externally observable from the viewpoint of time of speech, *kamoshire-nakat-ta* can express hypotheticality. The following sentences are examples that include *kamoshire-nakat-ta:*

16. More detailed discussion is found in chapter 5, where conditionals are investigated.

(26) a.

Anotoki	okane	ga	at-tara
at that time	money	NOM	have-COND
uchi	o	kat-ta	kamoshire-nakat-ta.
car	ACC	buy-PST	AUX-PST

"If I had money at that time I might have bought a house."

b.

Kyonen	sotsugyoo shite-i-tara,
last year	graduate-PERF-COND

imagoro	wa	moo	hataraite-i-ta	kamoshire-nakat-ta.
now	TOP	already	work-PROG-PST	AUX-PST

"If I had graduated last year, I might have been working by now."

The interpretation of sentence (26)a is either that it was possible that the speaker would have bought a house or that it was possible that s/he would not have bought a house in that situation. Although the reality concerning the actualization of the event in sentence (26)a is that the speaker did not buy the house due to a lack of money, this sentence expresses the indeterminability of a proposition—that is, the speaker still does not have a clear picture as to what might have happened. Sentence (26)b is open to the same interpretation: it is possible that the speaker would already be working if s/he had graduated, but the speaker him/herself does not know what would have happened. Therefore, *kamoshire-nakat-ta* can also be used to express hypotheticality, and choosing one form over the other depends on the speaker's point of view. However, even though the speaker in sentences (26)a and (26)b could have used *kamoshire-nai* instead of *kamoshire-nakat-ta,* the latter is often used in reference to a past situation that a speaker can no longer affect. In other words, *kamoshire-nakat-ta* is used as a remote form of *kamoshire-nai,* and when a speaker is in a situation where the possibility still exists for him/her to exert some form of control over the situation in question, *kamoshire-nakat-ta* is inappropriate. This means that although both *kamoshire-nai* and *kamoshire-nakat-ta* express hypotheticality, they are not necessarily interchangeable.[17] The difference between these two forms is illustrated in the following examples:

17. This was evident in example sentence (27)c, where the situation that speaker B in fact did not buy a car is clear.

(27) a. A: Okane ga at-tara kuruma o kau?

money NOM have-COND car ACC buy

"If you had money, would you buy a car?"

B: Kau kamoshire-nai.

buy AUX

"I may buy one."

* C: Kau kamoshire-naka-ta.

buy-PST AUX-PST

"It might have been the case that I will buy."

b. A: Okane ga at-tara kuruma o kat-ta?

money NOM have-COND car ACC buy-PST

"If you had money, would you have bought a car?"

B: Kat-ta kamoshire-nai.

buy-PST AUX-PST

"Yes I may have bought one."

? C: Kat-ta kamoshire-nakat-ta.

buy-PST AUX-PST

"I might have bought one."

c. A: Anotoki okane ga at-tara kuruma o kat-ta?

At that time money NOM have-COND car ACC buy-PST

"If you had money at that time, would you have bought a car?"

B: kat-ta kamoshire-nai.

Buy-PST AUX

"I might have bought one."

C: Kat-ta kamoshire-nakat-ta.

buy-PST AUX-PST

"I might have bought one."

In sentence (27)a, the speaker A is inquiring as to the future possibility of purchasing a car under the condition of having money, while in sentences (27)b and (27)c, the speakers are concerned with past situations. First, *kamoshire-nakat-ta* cannot be used when the situation described by the proposition is in nonpast context; accordingly, *kamoshire-nai* and *kamoshire-nakat-ta* are noninterchangeable in such a situation. Sentence (27)b is concerned with a past situation and both modals are interchangeable, although the utterance of speaker C in (27)b may be considered slightly awkward. This observation is perhaps due to the fact that the sentence still gives room for the circumstances to be manipulated through the interlocutor buying a car. This possibility is totally cancelled when the word *anotoki* "at that time" is added, as seen in the utterance of speaker C in sentence (27)c. Even though the sentence demonstrates an indeterminability of the speaker's decision making as to the purchase of the car, it creates a strong implication that the speaker actually did not buy a car and the speaker's purchasing a car cannot be affected by the current situation.

The use of *kamoshire-nai* in the past tense thus differs from the use of past tense *hazu* and *ni chigai-nai*. *Hazu-dat-ta* suggests to the interlocutor a counterfactual interpretation, and therefore it emphasizes the proposition's nonactualization. *Ni chigai-nakat-ta,* however, does not necessarily convey a counterfactual meaning. It implies that the speaker's judgment was made when an event occurred, and suggests that the speaker still believes his/her judgment concerning the truth of the proposition. *Daroo* does not partake in past tense forms, and most appropriately co-occurs with the adverb *tabun* "probably" or "perhaps." *Kamoshire-nakat-ta* expresses that the truth of a past event was indeterminable, although it does express that the speaker's current state is one counterfactual to that expressed in the original proposition.[18]

18. It is intriguing that *kamoshire-nai* and *kamoshire-nakat-ta* behave differently than the English auxiliaries "may" and "might." Coates (1983, p. 147) discuses the difference between "may" and "might" based on the idea that "might" is the tentative or unreal form of "may" and that they are usually interchangeable in their epistemic usage. She also quotes Wells (1979), Perkins (1981), and others who posit that child language studies indicate that "might" supersedes "may" as the main expression of epistemic possibility.

Thus, *kamoshire-nakat-ta* is used to hypothetically access a past event at the present moment.

It is clear that both *kamoshire-nai* and *kamoshire-nakat-ta* can emphasize the hypotheticality of an expressed proposition. The judgment leans toward neither realization nor nonrealization of the proposition, and *kamoshire-nai* and *kamoshire-nakat-ta* are not always interchangeable. The latter may be replaced with the former, but the converse is not possible since *kamoshire-nakat-ta* is used to indicate hypotheticality at a specific time in the past. Both forms deserve consideration as suppositionals that express epistemic possibility.

3.2.5 Summary of Suppositionals

The examination of the suppositionals revealed some characteristics and meanings of these modals. Although there are many other modals that appear in modal content (as seen in chart [14] in chapter 2), these are concerned with the notions of necessity, possibility, and the degree of speaker conviction regarding the truth of the proposition. The following provides a brief summary of each suppositional based on the examination given thus far.

Hazu

1. *Hazu* can function as either a suppositional or a nominal, exhibiting a speaker's strong belief that the proposition is true or realized (I'm sure that~) or a speaker's expectation that the proposition is true or realized (it is supposed to~), respectively.

2. *Hazu* functions as a suppositional only when used in nonpast tense form (*hazu-da*), and in such cases, the speaker's belief must be derived from objectively perceptible evidence (e.g., something personally witnessed or heard). Furthermore, the conditions at the time of speech involve a perceptual distance between the speaker and the acquired evidence.

3. Therefore, use of *hazu* to make judgment in the midst of an ongoing situation creates awkwardness, since such progressive situations cannot be viewed externally and with complete objectivity.

4. In a nonpast context, *hazu* may function as a nominal if the proposition is concerned with man-made principles, rules, plans, etc.

Such behavior at root communicates the same meaning as *-kotoni-natte-iru* "it has been decided/it is supposed to be."

5. *Hazu* functions as a nominal when used in negative or past tense forms (*hazu-ja-nai/hazu wa nai* and *hazu-dat-ta*), losing its function as a modal and expressing a meaning equivalent to *wake wa nai* "there is no way" or *-kotoni-natte-iru* "it was supposed to."
6. When used in past tense form (i.e., *hazu-dat-ta*), *hazu* creates a counterfactual interpretation indicating that the actual event is opposite to what is stated in the proposition.

Ni chigai-nai

1. Derived from the expression *machigai wa nai*, "there is no mistake," *ni chigai-nai* conveys a speaker's strong belief as to the truth of the proposition.
2. The speaker's judgment, however, does not have to be based on firm evidence but can simply express a strong belief founded on intuition as well as perceptible evidence.
3. Accordingly, *ni chigai-nai* is not used to express an uncertain, doubtful attitude on the part of the speaker
4. *Ni chigai-nai* can be used in past tense form (*ni chigai-nakat-ta*), but it may not quite objectify the past event. Since it does not necessarily indicate the actual outcome of the situation, the speaker may be still uncertain of the truth or realization of the proposition at the time of speech—that is, *ni chigai-nakat-ta* may function as a modal in past tense form.

Daroo

1. *Daroo* does not partake in negative form or past tense form.
2. *Daroo* can function either as confirmation of a proposition (equivalent to the sentence final particle *ne*) or as judgment of probability following a proposition.
3. When *daroo* is used as a tag marker, the sentence cannot include a question word. A sentence such as *Doo sure-ba ii daroo ne* is basically the same as *Doo sure-ba ii daroo ka* "What shall I do?" and gives

the polite equivalent of *desu ka*. The information described by a proposition is in the domain of the speaker's territory.

4. When used as a suppositional, *daroo* expresses a speaker's conjecture based on the environment in which the speaker delivers judgment, which in turn can be based on firm evidence or a speaker's intuition.

5. Suppositional *daroo* indicates a relatively wide range of implication, involving both necessity and possibility, in regard to the speaker's confidence in the truth or realization of the proposition.

6. The degree of speaker conviction is not always clearly specified by the presence of *daroo*; hence, *daroo* sentences are often taken to be noncommittal statements.

Kamoshire-nai

1. Literally meaning "it cannot be known," *kamoshire-nai* is used to express situations where the outcome cannot be determined by the speaker, that is, indeterminable situations. It neither implies that X will happen nor that X will not happen, but it is compatible with both.

2. *Kamoshire-nai* is used to express that the proposition has neither truth value nor determinability; therefore, *kamoshire-nai* expresses hypotheticality.

3. The past tense form, *kamoshire-nakat-ta*, also exhibits the indeterminability of a situation described by the proposition: the speaker still does not know at the time of speech whether the proposition was realized or not.

4. Both *kamoshire-nai* and *kamoshire-nakat-ta* therefore function as modals and express a low degree of speaker conviction, that is, a high degree of modality.

Among suppositionals, therefore, the degree of speaker confidence in the truth of a proposition is expressed most highly by *hazu*, followed by *ni chigai-nai*, *daroo*, and to the lowest extent by *kamoshire-nai*. This semantic result should correspond well with the intuitive judgment of native speakers of Japanese. Incorporation of evidentials into the continuum of the degree of speaker conviction requires further investigation, as given in the following sections.

3.3 EPISTEMIC MODALITY—EVIDENTIALS

According to Palmer (1986), there are many languages in which the epistemic system appears to consist of both evidentials and judgments: evidentials are constituents that refer to visual or sensory evidence, and judgments refer to modal auxiliaries that are used to express a speaker's supposition. He cited Barnes, who referred to Tuyuca as an example of a pure evidential system. According to Barnes, there are five ways in which information is obtained: "visually; through a sense other than the visual; through evidence of the state or event; by being told about the state or event; or by assuming what happened, and there is a hierarchical relation between these terms" (1984, p. 255). Barnes further states that the visuals are the reliable evidentials and that they are used when a speaker has witnessed, or is presently observing, a situation or event. Tuyuca is basically an SOV word order language with the evidential forms following the verb. The following sentences are examples in Tuyuca illustrated by Barnes (1984):

(28) a.

Díiga	apé-	wi
he	play	evidential

"He played soccer. (I saw him play.)"

b.

Díiga	apé-	ti
he	play	evidential

"He played soccer. (I heard the game and him, but I did not see it or him.)"

c.

Díiga	apé-	yi
he	play	evidential

"He played soccer. (I have seen evidence that he played: his distinctive shoe print on the playing fields. But I did not see him play.)"

d. Díiga apé- yigi

he play evidential

"He played soccer. (I obtained the information from someone else.)"

e. Díiga apé- hiyi

he play evidential

"He played soccer. (It is reasonable to assume that he played soccer.)"

All five examples convey the same proposition ("He played soccer"), but all are nondirect forms by virtue of the evidentials *wi, ti, yi, yigi,* and, *hiyi,* which communicate the means by which the speaker obtained the information Barnes proposed five descriptors for these evidentials: visual, nonvisual, apparent, secondhand, and assumed, respectively. Although Barnes used the term "evidential" (and not "auxiliary"), these evidential morphemes are structurally quite similar to the Japanese modals *soo, yoo,* and *rashii* (referred to as "evidentials" hereafter), in that each follows a proposition and indicates how a speaker obtained the information.

The distinction between a proposition and a proposition with an evidential corresponds to that between direct and nondirect forms, respectively, in Kamio's theory of territory of information (1990, 1997b) as discussed in section 3.2.3. These forms are concerned with territories inside and outside the speaker's domain, respectively, and correspond intimately with the notion of "evidentiality." The use of direct forms indicates that the information conveyed by the sentence is true (to the best of the speaker's knowledge) and belongs to the speaker's territory. On the other hand, nondirect forms signify first of all that it is not within the speaker's power to establish the truth of the proposition, and secondly, either the information conveyed by the sentence does not belong solely to his/her territory, or does not belong to his/her territory at all.

Kamio also relates the notion of evidentiality (nonbare-form expressions) to the theory of mental spaces proposed by Fauconnier (1994). Mental spaces are constructs distinct from linguistic structures but built up in any discourse according to guidelines provided by the linguistic expressions, which are referred to as *space-builder* expressions.[19] Nonbare-form expressions are identical to space-builders that establish the psychological spaces corresponding to an actual world and a world that a speaker creates through his/her imagination (i.e., an imaginary world). For example, if a speaker says, "I hear that Mary is pregnant," there are two worlds implied: the world in which she is actually pregnant and another based on what the speaker hears. This is exemplified in the following illustration:

(29)

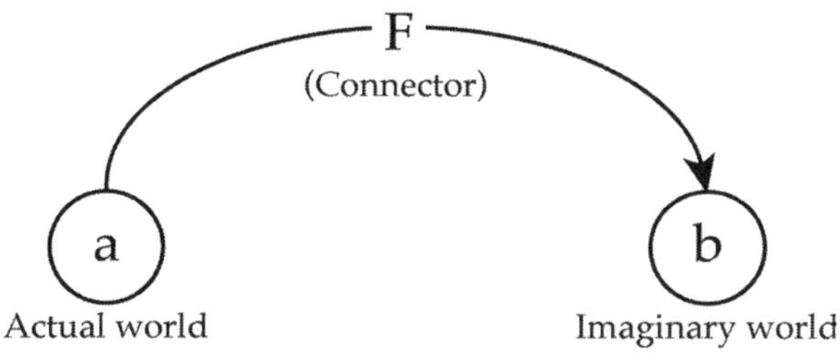

Comparing the theory of territory of information to Fauconnier's theory, we see that the inside and outside territories of Kamio correspond to actual worlds and imaginative worlds, respectively. Evidentials function as space-builders, which in turn shows that space-builders are deeply related to modality as well. The following are a number of sentences that represent Japanese evidentials:

(30)	Nihonkai	no	sakana	wa	totemo	oishii	soo-da.
	Japan sea	GEN	fish	TOP	very	delicious	AUX-COP

"I hear that fish caught in the Japan Sea are very delicious."

(31)	Kyoo	no	sakana	wa	totemo	oishi-soo-da.
	today	GEN	fish	TOP	very	delicious-AUX-COP

"Today's fish appears to be very delicious."

19. According to Fauconnier, space-builder expressions may be prepositional phrases, adverbs, connectives, underlying subject-verb combinations, and so forth and are not limited to evidentials.

(32) Jon ga daihyoo ni erab-are-ta yoo-da.
John NOM representative as elected-PASS-PST AUX-COP
"It looks like John was selected as the representative person."

(33) Kabu no hendoo ni tomonai, en ga kyuu jooshoo shi-ta rashii.
stock GEN change along with yen NOM sudden rise-PST AUX
"Along with the change in the stock market, it seems that the yen experienced a sudden increase."

As seen in sentences (30) and (31), *soo* expresses two different meanings depending on the form of the preceding word. While *soo* following the gerundive form expresses a judgment based on a visual impression, *soo* following the tense form expresses secondhand hearsay information. Hereafter, these are distinguished in the following manner: *soo* following the stem/gerundive form is represented by "visual *soo*," and *soo* following the tense form is referred to as "hearsay *soo*."

In the use of these Japanese evidentials, it appears that two elements—information obtained from an outside source, and visual and/or sensory-based judgment—are subtly interconnected. When asked for information on any specific matter, if lacking in direct experience regarding the matter, the speaker makes inferences based on information which is acquired from related experience. Speakers can also act merely as conveyers of information from an outside source, communicating information without processing it. Evidentials also express modality in that they are all concerned with a speaker's supposition about the possibility of the expressed proposition. A brief examination of discrete evidentials is given below.

3.3.1 *Soo* (hearsay)

Soo, when preceded by a perfective form, can co-occur with all predicate forms. It is fixed in form and does not partake in a negative, past, or interrogative form. It follows a proposition, expressing the equivalent of "that is what I heard," and is used to convey secondhand information to another party. It is used when the speaker wants to signal to the interlocutor that the information represented by the proposition has been obtained from a secondary source, and that s/he is not responsible for the truthfulness or falsehood of the information. Therefore, when a speaker wishes to make the method of obtaining information clear or to quote another person directly, *~to it-ta* "(someone) said" or *~kara ~to kii-ta* "I heard it from ~" is used as a direct link to some past discourse. Note that direct and indirect discourse are structurally distinguished in English,[20] and the difference between these discourses is much clearer than it is in Japanese. The grammatical markings of person, tense, and time differ in indirect and direct discourse in English, whereas such distinctions are not clearly marked in Japanese. The following sentences compare the use of *soo* and the verb *kii-ta:*

(34)	Kare	wa	kekkonshi-nai	soo-da.
	he	TOP	marry-NEG	AUX-COP

"I heard that he would not get married."

(35)	Watashi	wa	kare	wa	kekkonshi-nai	to	kii-ta
	I	TOP	he	CNT	marry-NEG	QUOT	hera-PST

"I heard that he would not get married."

Although it is not clear from these sentences alone whether the source of information is a particular conversation or a direct discourse in which the speaker participated, sentences (34) and (35) are semantically identical in conveying information that has been heard or read, regardless of where it was heard or read, sharing one common point that the speaker has obtained the information that *Kare wa kekkonshi-nai*, "He will not get married." Neither sentence involves a speaker's conjectural judgment. The speaker does not have to take responsibility for what is said, since statements ending in *soo* are not based on any of his/her own assumptions. There is no involvement of visual or sensory impression implied in *soo* sentences;

20. For example, he said that he would not come is an indirect quotation that corresponds to the direct quotation he said, "I will not come." In Japanese both structures are represented by *Kare wa konai to it-ta,* for which quotation marks are often omitted.

instead, the speaker is simply delivering outside information directly to an interlocutor.

As Kamio (1990, 1997b) observed, even information obtained from reliable sources, such as newspapers, radio and television news, and weather forecasts, is conventionally treated as "hearsay" information because it has been obtained from a secondary source.[21] In daily conversation *soo* sentences are often used to indicate the basis for the judgment of the situation a speaker is in—that is, the speaker believes what s/he hears to be true and relies on the information when giving his/her opinion. Note the following dialogues:

(36) A: E, ya-da! Tomo ga ashita taifuu sekkin-da-tte.
oh no Tomo GEN tomorrow typhoon approach-COP-QUOT
"Oh, no! Tomo says the typhoon is approaching us tomorrow."

B: E, soo? Nihonkai gawa ni nukeru-tte kii-ta kedo?
oh really Japan seaside LOC go across-QUOT hear-PST but
"Really? I heard that it will go across the Japan seaside, but . . . (Isn't that right?)"

C: Soo soo, nihonkai gawa ni nukeru soo yo.
that's right Japan seaside LOC go across AUX VOC
"That's right. I hear that (the typhoon) will go across the Japan seaside."

21. There are cases where such information can be treated as directly obtained information. For example:

(a) Norikumi in hyakujuuhahchi nin no kyuushutsu wa
crew 118 people GEN rescue TOP
kiwamete konnan-da soo-da.
extremely difficult AUX-COP
"The rescue of the 118 crew members will be extremely difficult, I hear."

Tookyoo no kion wa kyoo no gogo
Tokyo GEN temperature TOP today GEN afternoon
sanjuuhachido made agaru soo-da.
degree up to rise AUX-COP
"Today's afternoon temperature in Tokyo will go up to 38 degrees, I hear."

Dakara	daijoobu.	Hikooki,	tobu	yo,	zettai.
so	do not worry	airplane	fly	VOC	definitely

"So, don't worry. The airplane will depart, definitely."

A:

Fuu,	yokat-taa.
Whew,	good-PST

"Whew, I'm relieved to hear that."

(37) A:

Rainen	kochira	e	irassharu	soo-desu	nee.
next year	here	LOC	come	AUX-COP	PART

"I hear that you are coming here next year."

Ee.	Doozo	yoroshiku	onegai shimasu.
yes	please	good	do

"Yes. Please extend your favor to me (I hope we can be good friends)."

B:

Iie,	kochira koso
no	same to you

"Same to you."

In dialogue (36), speaker C (clearly taking for granted what s/he has heard) assures speaker A that the airplane will leave. There is no proof for speaker C's statement except his/her belief that the information came from a reliable source. The same can be observed in dialogue (37), where speaker A's strong belief that speaker B is coming to town is evident in the exchange. Both speakers are relying on the truth of the obtained information to develop the dialogues. An accurate interpretation of *soo* is, therefore, "I heard, so I know/understand," which expresses a very high degree of necessity.

One interesting behavior of *soo,* however, is that it can follow another modal, especially suppositionals. This characteristic of modals is examined in the section concerning compatibility tests (3.4.3), but some concrete examples are helpful at this point:

(38) 3ji ni kaigi ga aru hazu-da/kamoshire-nai soo-da.
3 o'clock TEMP meeting NOM exist AUX-COP/AUX AUX-COP
"I hear that there should be a meeting at three o'clock."

(39) ?? a. Tatsumaki ga kuru yoo-da/rashii soo-da.
tornado NOM come AUX-COP/AUX AUX-COP
"I hear that it looks like/seems like a tornado is coming."

b. Tatsumaki ga kuru yoo-dat-ta/rashikat-ta soo-da.
tornado NOM come AUX-COP-PST/AUX-PST AUX-COP
"I hear that it looked/seemed like a tornado was coming."

c. Tatsumaki ga kuru yoo-da/rashii tte kii-ta.
tornado NOM come AUX-COP/AUX QUOT hear-PST
"I hear that it looks like/seems like a tornado is coming."

When *soo* is used with the suppositionals *kamoshire-nai* and *hazu* it automatically lets the preceding modals function as indicating another individual's uncertain thoughts or judgment. *Hazu* and *kamoshire-nai* in sentence (38)a are representing someone else's speculation regarding the existence of a meeting, and the speaker is conveying that information. The suppositionals *hazu* and *kamoshire-nai,* and the evidential *soo* have a different role in this sentence; only *soo* takes the speaker as a covert semantic subject.

Judgment based on auditory/visual impression processed by other individuals, however, is difficult to convey at the time when such judgment is being made and has not yet become objectively perceptible. The awkwardness exhibited in sentence (39)a is attributed to combining what the speaker hears and some judgment made by another individual simultaneously at the time of speech. This awkwardness can be removed when the evidentials *yoo* and *rashii* are used in past tense forms, or when *soo* is replaced by the verb *kii-ta* "I heard." This is exemplified in sentences (39)b and (39)c, respectively. *Yoo-dat-ta* in (39)b means that a tornado seemed to come at one point in the past, but apparently it did not come. The speaker is conveying the individual's judgment that such an occasion occurred in the past. The semantic representation of sentence (39)c is identical to that of (39)a, yet the degree of appropriateness is higher than that of sentence (39)a. This difference is attributed to the use of the past tense form of the verb *kiku,* "to hear," which lets us perceive a point where the speaker obtained the information. Thus, although *soo* is semantically equivalent to *kii-ta* "I heard," a subtle difference between them is recognized by comparing sentences that co-occur with another evidential. This observation confirms that *soo* is not simply an equivalent of *kii-ta,* which is a constituent of propositional content, and that *soo* as an evidential should be treated as one of the modals, expressing a high degree of epistemic necessity in that the speaker believes/knows the truth of the proposition.

3.3.2 *Soo* (appears to be)

A second *soo,* originating from the nominal *yoosoo,* refers to a state of affairs that is externally apparent (Morita 1980, p. 220) and can exhibit a concrete meaning by itself. This *soo* is referred to as visual *soo,* which is attached to the stem form of adjectives,[22] nominal-adjectival, and the gerund form of verbs. Due to its syntactic dependency, a clear line distinguishing modal content from propositional content cannot be drawn. Yet, as Bybee (1995) states that modality is in the semantic domain, it is the meaning of modals that is the primary consideration, and visual *soo* is considered accordingly among the evidentials.

Visual *soo* can partake in negative, past tense forms and it can also be used in questions, although when *soo* is used with the negative form of a verb it occurs primarily as either *soo-ni-nai, soo-mo -nai,* or *nasa-soo.*[23] Which one is used in day-to-day conversation is decided

22. Exceptions to this are that *nai* becomes *nasa-soo,* and *ii* becomes *yosa-soo.*

23. When a sentence is in negative context, either *soo-ni nai, soo-mo nai,* or *nasa-soo-da* is used. For example, *kirei-ja-nai* "not pretty" can form *kirei-ja-nasa-soo* "does not look pretty" in the same fashion as the adjective *kitanai* "dirty" forms *kitanasa-soo* "looks dirty." When a verb is negated, however, the verb becomes the primary focus. For example, when negating the verb *furu* "to fall (rain)" the primary choice is *furi-soo mo nai* "it appears that it will not rain." The verb form *ame wa fur-anasa-soo* is a rather colloquial speech form and is seldom seen in writing.

by personal preference. Visual *soo* represents the speaker's observation cotemporal with the time to which *soo da/dat-ta* refers.[24] The following are examples in which *soo* partakes in the past tense form:

(40) Arashi wa Tookyoo e mo ki-soo-dat-ta.
storm TOP Tokyo LOC too come-AUX-COP-PST
"The storm almost hit Tokyo too."

(41) Kare wa gaaru furendo ni fur-arete, kanashi-soo-dat-ta.
he TOP girlfriend by break up-PASS sad-AUX-COP-PST
"He broke up with his girlfriend, and he looked sad."

(42) Kaasan, kaze hiite-i-soo-dat-ta yo.
mother cold catch-PERF-AUX-COP-PST VOC
"Mother appeared to have caught a cold."

In considering sentences (40)–(42), it is intriguing to note that there is a difference detected in terms of the possible occurrence of the event. Sentence (40) clearly indicates that the situation described in the proposition seemed as though it was about to be realized and expresses a sense of immediacy, but in reality, such was not the case at the time when the sentence was uttered. On the other hand, sentences (41) and (42) do not imply such counterfactual meaning, but express that the truth of the proposition is unknown, and each simply indicates the speaker's belief based on a visual impression of the situation.

24. It should, however, be mentioned that there are cases where visual *soo* is used without an actual observation of things in question. Please note the following examples:

(a) Kyoo no bangohan wa ne, tonkatsu, agedashi-doofu,
today GEN dinner TOP PART pork cutlet fried tofu
sorekara nasu no gomaae, sorekara . . .
also egg plant GEN dressed with sesame sauce then
"Today's dinner includes dishes such as fried tofu with sweet sauce, cooked eggplant dressed with sesame sauce, in addition to pork cutlet, and then . . ."

(b) Waa, oishi-soo!
wow delicious-AUX
"Wow, it sounds delicious."

As the English equivalent indicates, the visual *soo* used here refers to delicious sounding food. Although "it sounds" is the English equivalent, the speaker is visualizing the dinner as the listener describes it, and therefore this use of *soo* is still based on what the speaker sees in the imaginary picture at the time of speech.

The different interpretations may be approached in terms of the type of predicate employed in each sentence: sentence (40) contains a nonstative predicate,[25] while sentences (41) and (42) each contain a stative predicate—respectively, the adjectives "sad" and the verb "to catch (a cold)" accompanied by the *te-iru* form expressing a perfect meaning.[26] Nonstative predicates tend to force *soo* sentences to express that the event was about to be realized, but failed, whereas stative predicates tend to force *soo* sentences to express that the truth of the proposition cannot be determined; i.e, the speaker is less certain concerning the truth of the proposition. Such an observation suggests that there is a deep correlation between stative predicates and modality, which is evident in the present tense use of *soo* as well. The following examples involving nonstative predicates illustrate this point:

(43)	A,	biiru	ga	kobore-soo-da	yo!
	ah	beer	NOM	spill-AUX-COP	VOC

"Watch out! It looks like the beer is about to spill."

25. The term "stative" here is employed generally on the basis of Comire's definition of "stativity," which refers to a pure state that does not require energy input to maintain the state; it continues eternally unless some kind of outside energy is added. This description also communicates the characteristics of stative predicates proposed by Soga (1983, p. 88), having 13 semantic characteristics, with the common semantic feature of stative predicates being the feature of nonself-controllability. In Japanese, stativity is expressed in the form of adjectives, nouns, and stative verbs such as verbs of existence, ability, spontaneity, and other verbal predicates that cannot offer the concept of "activity," such as *iru* "to need" and *nite-iru* "to resemble."

26. It should be mentioned, however, that there is a degree to which stativity is expressed in a predicate. Some compound verbs such as potential and passive forms may have a high possibility of conveying the meaning "about to be~." Please note the following examples:

(a) Moo sukoshide	A	ga	mor-ae-soo-dat-ta.	
a little more	A	GEM	obtain-POT-AUX-COP-PAST	

"I almost got an A grade."

(b) Okane	wa	hikidashi	ni	ari-soo-dat-ta
money	TOP	drawer	LOC	exist-AUX-COP-PAST

"It looked like the money was in the drawer."

Although *mor-a-eru* "to be able to obtain" is a compound verb with a potential suffix and is considered a stative predicate, it can be interpreted as emphasizing the meaning of either "I was about to receive an A" or "it looked like I was able to receive an A" depending on the situation. In this case, however, the former interpretation may prevail over the latter since the actual outcome of the event is most likely that the speaker did not receive an A grade. *Aru* in sentence (b) is also a stative verb, yet, this does not render the "about to be~" interpretation when it is used with *soo*. Thus, the difference in degree is detected. The interpretation depends on whether the original verb or the feature of stativity suffixed to the verb dominates the meaning of the whole predicate.

(44)

Kono yoosu-ja,	moosugu	ame	ga	furi-soo-da.
this appearance-COP	soon	rain	NOM	fall-AUX-COP

"Based on the appearance of the sky, it looks that it will rain soon."

(45)

Neko	ga	teeburu	kara	ochi-soo-ni	natte-iru.
cat	NOM	table	ABL	fall-AUX	become-PERF

"My cat is about to fall off the table."

In a nonpast context, when *soo* is used with a nonstative predicate, it expresses a judgment that something is about to happen in the near future and in the presence of the speaker. This is due to the fact that nonstative predicates indicate future tense or habitual meaning when used in a citation form. The speaker's assertion is judged according to the situation the speaker is currently observing and suggests the change of the event at any moment with a sense of immediacy. For this reason, a *soo* form with a nonstative predicate is compatible with the adverb *imanimo* "at any moment." Additional characteristics of *soo* may also be seen in the following examples:

(46)

Onaka	ga	sui-te	imanimo	shini-soo-da.
stomach	NOM	empty-GER	at any moment	die-AUX-COP

"I'm so hungry I feel like I'm going to die at any moment."

(47)

Shiken	no	koto	o	kangaeru-to	imanimo
test	GEN	about	ACC	think-COND	at any moment

ki	ga	kurui-soo	ni-naru.
mind	NOM	crazy-AUX	become

"When I think about the test I feel like I'm going to go crazy at any moment."

(48)

Me	ga	mawat-te	imanimo	taore-soo-da.
eyes	NOM	turn-GER	at any moment	pass out-AUX-COP

"I'm dizzy and feel like I'm going to pass out at any moment."

Sentences (46)–(48) reveal some characteristics not observed so far in that they are not based on a direct observation of the situations; rather, they imply an indirect visual observation with the speaker being both subject and object of the observation. They refer to an emotional or mental state that the speaker almost, but not quite, achieves. Sentences (46)–(48) all express that something is about to happen, and all contain a nonstative predicate with the adverb *imanimo* "at any moment." They emphasize that the content of the proposition is very close to being realized based on the speaker's visual observation or information from his/her senses. The speaker is in effect processing information gained from his/her own objective view of him/herself. A stative predicate, which does not imply a change of state, therefore cannot be used to express these situations.

Consider now the use of *soo* in a nonpast context with stative predicates. A stative predicate used alone can express present or future situations. When *soo* is added, the sentence expresses the speaker's visual or sensory impressions, as shown in the following examples:

(49)	Sono	seetaa	taka-soo	ne.
	that	sweater	expensive-AUX	PART

"The sweater you are wearing looks expensive."

(50)	Konna	tokoro	nara	yuuree	demo	i-soo-da.
	such	place	COND	ghost	or something	exist-AUX-COP

"It looks like there may be a ghost or something in a place like this."

(51)	Akubi	bakari	shite-i-te	ikanimo	hima-soo-da.
	yarn	only	do-PROG-GER	really	bored-AUX-COP

"He is doing nothing but yawning and looks like he is really bored."

Soo in these examples is used to describe a visual impression. All of the above sentences can co-occur with the adverb *ikanimo* "almost as if." The expression in sentence (49) *taka-soo* "looks expensive" combines the present state of the sweater and the speaker's conveying a visual impression received from the sweater. It does not matter whether the sweater is actually expensive or not, nor does it matter what it looks like to other people. The sentence merely indicates the speaker's belief that the sweater looks expensive to the speaker. In these examples, the speaker is making a judgment based on visual impressions of the situation s/he is observing at the time of speech. For example, in (50), based on the darkness and desolateness of the place, the speaker makes the judgment: "it looks like there is a ghost." This, however, is not necessarily an accurate conclusion; it may be true or may not be true. In that sense *soo* has a close affinity to the function of the suppositional *kamoshire-nai*. It must be noted, however, that the visual evidence upon which this form is based creates the supposition of a higher likelihood of occurrence than is the case with *kamoshire-nai*.

Observation of the behavior of past and nonpast tense *soo* with different types of predicate thus far leads one to the same conclusion: nonstative predicates co-occurring with *soo* express the imminent realization of a proposition, while stative predicates with *soo* express a state of affairs together with the speaker's observations based on a sensory-based impression received at the time of speech. When *soo* is used with a nonstative predicate, the situation the speaker is/was observing could change at any moment, and the use of *soo* creates a sentence that conveys a snapshot of the last moment of some action, before the inevitable conclusion of that action. When *soo* is used with a stative predicate, on the other hand, the appearance of the object is/was trapped in the moment the speaker is/was in, however, and there is/was no implication of the immediate occurrence of later change. Therefore, the possibility of the proposition's truthfulness may be less than that of *soo* with a nonstative predicate. What is observed here is that stativity, which does not involve change, may create a spatiotemporal distance between the speaker and the situation and be farther away from reality; hence, it is closer to the notion of reduced possibility and nonactuality.[27]

27. As noted by Kuno (personal communication 2001), it is a tendency and not an absolute requirement that stative predicates create a spatiotemporal distance between the speaker and the situation. For example, when a person sees fish at a market and *shouts Waa, kyoo wa sakana ga oishi-soo-da!* "Wow, the fish looks delicious today!" there is no implication that the proposition is far away from the reality. The speaker is actually in front of the fish and making a judgment. The point in this discussion is to detect the minute difference created by the type of predicate in terms of the actuality/reality of the situation. A difference may also be seen in the following two sentences:

(a) Kondo no shiken wa muzukashi-soo-da.
this time GEN exam TOP difficult-AUX-COP
"I feel that the coming exam is difficult."

(b) Wa, kondo no shiken, muzukashi-sooda.
Wow this time GEN exam difficult-AUX-COP.
"Wow, this test looks difficult."

It is clear that the predicate type has a significant impact on the interpretation of visual *soo* sentences. In any case, the point to be emphasized is that these meanings all involve human judgments based on either a situation that has already been altered into a certain state as a result of prior change or pure state that does not involve change, or is about to be realized, both of which are close to the truth from the speaker's viewpoint and do not invite speaker inference. This temporal quality is also crucial in determining the degree of modality. Such a feature can further be authenticated through examination of *rashii* and *yoo,* each of which involves a greater temporal distance and a higher degree of speaker inference than *soo.*

3.3.3 *Rashii* (seems like)

Rashii is used with the plain form of verbs and adjectives, either in the nonpast or past tense. When the proposition involved is in nonpast context, *rashii* immediately follows a nominal or a nominal-adjectival without allowing a tense marker between the propositional content and itself. *Rashii* is adjectival in nature and inflects accordingly, but it cannot be used in interrogative or negative forms.

Rashii is used to represent a speaker's a presumptive judgment that is derived from outside information.[28] Morita (1980) and Teramura (1984) state that *rashii* and *yoo* are often interchangeable without affecting the meaning of the sentence, but the difference between these two evidentials is that *rashii* is used when a speaker

---

The difference between sentence (a) and (b) is the physical distance of the exam from the speaker: it is best interpreted that the speaker is not actually looking at the exam in sentence (a), while sentence (b) is uttered as soon as the speaker takes a look at the exam as illustrated by English equivalent. Although the speaker believes that the proposition is true for both sentence (a) and (b), a situation that is about to alter into a different state, such as *A, ochi-soo-da!* "Watch out! it's going to fall!", change of the state that is actually occurring in front of the speaker or in his/her mind might be more convincing that such alteration shows higher degree of being true.

28. *Rashii* has one more meaning not considered here - that which results when it is used with a noun to mean "ideal." This meaning can be made clearer when *ikanimo* "almost as if" is added to the sentence, but the meaning is usually understood from context. This *rashii* behaves the same way as an adjective except that it never appears in the past tense. Note the following examples:

(a) Otona rashii otona.
adult AUX adult
"An adult who behaves as an adult."

(b) Ikanimo sensei rashii hito.
almost as if teacher AUX person.
"He embodies as if he is a teacher."

makes a judgment based on a fact that is observable to the speaker, while *yoo* is used when a speaker relies on his/her own judgment. Both Morita and Teramura use the terms "objective" and "subjective" judgment in order to describe the meaning of *rashii* and *yoo*, respectively. Hayatsu (1998) also makes a similar remark, although she points out the equivocal use of these terms and avoids using them. Her examination concludes that *rashii* is used when a speaker perceives a situation that lies outside of the speaker's domain, and *yoo* is used when a situation is perceived that lies inside of the speaker's domain. This is identical to Kamio's (1990, 1997b) earlier observation that *rashii* being the essential notion of the territory of information. *Rashii* is an indirect expression that is employed based on uncertain evidence. The following are example sentences that compare *rashii* with *yoo:*

(52) a.

Atarashii	gakubuchoo	ga	kimat-ta	rashii.
new	department chairman	NOM	select-PST	AUX

"It seems like the new department chairman was selected."

b.

Atarashii	gakubuchoo	ga	kimat-ta	yoo-da.
new	department chairman	NOM	select-PST	AUX-COP

"It looks like the new department chairman was selected."

As many researchers state, *yoo* and *rashii* are generally interchangeable: the sentences do not become clumsy or grammatically incorrect if *rashii* is substituted for *yoo*, nor is the sense of the sentence destroyed. Nevertheless, as more information is added to the sentences, they are no longer quite as flexible. The following examples illustrate this point:

(53) a.

Chichi	no	hanashi	ni yoruto,	sensoo	chuu	wa
father	GEN	story	according to	war	during	TOP

imo	ga	naniyori	no	gochisoo-dat-ta	rashii.
potato	NOM	best	GEN	feast-COP-PST	AUX

"According to my father's story, it seems that potatoes were the best food during the war."

? b.	Chichi	no	hanashi	ni-yoruto,	sensoo	chuu	wa
	father	GEN	story	according to	war	during	TOP

imo	ga	ichiban	no	gochisoo-dat-ta	yoo-da.
potato	NOM	best	GEN	feast-COP-PST	AUX-COP

"According to my father's story, it looks like potatoes were the best food during the war."

(54) ?? a.	Nishi	no	sora	ga	akai	tokoro	o	miru-to,
	west	GEN	sky	NOM	red	appearance	ACC	look-COND

ashita	mo	hareru	rashii.
Tomorrow	also	clear	AUX

"From the appearance of the sky in the west, it seems that it is going to be sunny tomorrow too."

b.	Nishi	no	sora	ga	akai	tokoro	o	miru-to,
	west	GEN	sky	NOM	red	appearance	ACC	look-COND

ashita	mo	hareru	yoo-da.
Tomorrow	also	clear	AUX-COP

"From the appearance of the sky in the west, it looks like it is going to be sunny tomorrow too."

The fact that (53)a is more appropriate than (53)b suggests that *rashii* is the better choice when the source of the information lies outside the speaker. On the other hand, *yoo* becomes the more appropriate choice in sentences (54)a and b since *akai tokoro o miru-to*, "looking at the sky," indicates the speaker's present situation and s/he is making a judgment based on the appearance of the sky. When the information comes from outside the speaker, *rashii* is the preferred choice, and when the information comes from the speaker's direct perception, *yoo* seems to make more sense. This observation suggests that *rashii* and *yoo* are distinguished according to whether the judgment has been made based to a higher degree on outside information or to a higher degree on one's own intuition. In this sense, the use of *rashii* is closer to hearsay *soo* than *yoo* is in that *rashii* is used when an utterance has its roots in obtained information rather than a speaker's own observation of present appearance. In the use of *rashii,* information from the outside and a speaker's own supposition are somehow entwined.

Although previous research concerning evidentials has emphasized the similarity of *yoo* and *rashii,* detecting the difference may clarify the use of these evidentials. The following examples include cases that allow *yoo,* but not *rashii:*

(55) Standing in front of a mirror at a department store:

A: Nee, doo? Kono fuku niau kashira.
say how this dress suit wonder
"What do you think? Do you think it looks good on me?"

B: Un . . . Chotto aka-sugiru yoo/*rashii ne.
well a little red-too AUX/*AUX PART
"Well . . . it looks a little too red for you, I guess."

(56) A: Ano hito anna tokoro de nani shite-(i)ru-no kashira.
that person such place LOC what do-(PROG)-COMP wonder
"I wonder what that person is doing over there."

B: Odot-te(i)-ru yoo/*rashii yo.
dance-PROG AUX/*AUX VOC
"It looks like he is dancing."

A: Ya-da! Chigau yo. Tasukete-tte sakende-(i)ru-no yo!
no-COP wrong PART help-QUOT scream-PROG-COMP VOC
"No! That's wrong. He is screaming for help!"

(57) A: Ashita hareru kashira.
tomorrow sunny wonder
"I wonder if it's going to be sunny tomorrow."

B: Hareru rashii yo.
sunny AUX VOC
"It looks/seems like it's going to be clear."

C: Hareru yoo yo.
sunny AUX VOC
"It looks/seems like it's going to be clear."

While sentences (55) and (56) describe situations that are taking place in front of the speakers and *rashii* is inappropriate, sentence (57) is concerned with a future event where *rashii* and *yoo* are substitutable. There is, however, a subtle difference in the implications associated with the *rashii* and *yoo* versions. Speaker B in sentence (57) makes clear through the use of *rashii* that the source of the information lies outside the speaker—that is, s/he must have obtained the information from a weather forecast or such. In contrast, speaker C in sentence (57) implies through the use of *yoo* that the speaker has probably obtained the information, for example, by looking at the sky for some sort of sign. (The sentence can also be used when the speaker has obtained the information from an outside source.) Nevertheless, both *yoo* and *rashii* are used when a speaker does not know the truth of the proposition and therefore makes a suppositional judgment based on visual/perceptual or sensory/perceptual impression, respectively. They are both in the domain of epistemic possibility, locating *rashii* closer to the use of hearsay *soo*. The difference between *rashii* and *yoo* will be further clarified in the next section.

3.3.4 *Yoo* (looks like)

As with visual *soo,* the modal *yoo* is derived from the noun, *yoosoo,* referring to a state of affairs that is externally apparent. As Morita (1980, p. 503) shows, *yoo* means "the way things are," and it is important to note that the manner in which things are perceived is not solely based on visual impression, but on various sensations and perceptions. In this sense, *yoo* should be used in a context wider than that of visual *soo.*

Yoo may be added to verbs, adjectives, nominals followed by *no*, and nominal-adjectivals followed by *na. Yoo* can occur in past tense and question forms, but not negative forms. Teramura (1984, p. 243) explained that the fundamental meaning of *yoo* is "close to the truth," but this can be divided into two meanings: suppositional judgment and metaphor as in, for example, *Tatsumakin ga kuru yoo-da*, "It looks like a tornado is coming," and *Kare wa sakana-no yoo-ni oyogu* "He swims as if he were a fish," respectively. Teramura claimed that the first meaning comes about when a situation is such that a speaker is not sure if a proposition is true or not, but supposes that it may be very close to the truth based on his/her own observation. In this case the meaning of *yoo* is that of suppositional judgment. The latter meaning is carried when a speaker knows the truth value of the proposition, but recognizes that some other object(s) have an appearance that is very similar to the subject of the proposition. In this case *yoo* creates a "metaphorical" meaning. The objective in this section is to examine suppositional judgment as expressed in *yoo.*[29]

Yoo frequently co-occurs with the adverb *nandaka* "somehow," which is used when a speaker is not sure why s/he feels that way. *Yoo* also often co-occurs with *ki ga suru* "I feel like" that composes a set phrase and expresses a speaker's suppositional judgment based on intuitive perception. The following sentences exemplify these phrases with *yoo* and also compare the use of them with *rashii:*

29. As Teramura (1984) pointed out, the meaning of *yoo* is sometimes vague unless it is accompanied by adverbs such as *dooyara* "seems likely," *marude* "as if," and so forth. For example, a sentence *Ikita sakana no yoo-da* can be interpreted as either "It looks like a live fish," which expresses a speaker's supposition, or "It is as if it were a live fish," which expresses a metaphorical meaning. For the former, the speaker does not know the truth of the proposition (*iki-ta sakana*), while in the latter, the speaker knows the truth, that is, that the object is not a live fish, but it appears as such. The adverbs *dooyara* "perhaps" and *marude* "as if" can differentiate the meaning:

(a) Dooyara iki-ta sakana no yoo-da.
seems likely live-PERF fish GEN AUX-COP
"It looks like a live fish."

(b) Marude iki-ta sakana no yoo-da.
as if live-PERF fish GEN AUX-COP
"It looks as if it were a live fish."

(58) a. Imooto wa nandaka saikin yase-ta yoo-da.
younger sister TOP somehow recently lost weight-PST AUX-COP
"My sister looks like she has lost weight lately."

b. Nandaka jishin ga kuru yoo-na ki ga suru.
somehow earthquake NOM come AUX-COP feel like
"I feel like there may be an earthquake."

?? c. Imooto wa nandaka saikin yase-ta rashii.
younger sister TOP somehow recently lost weight-PST AUX-COP
"My sister seems like she has lost weight lately."

* d. Nandaka jishin ga kuru rashii ki ga suru.
somehow earthquake NOM come AUX feel like
"I feel like there seems to be an earthquake."

Sentence (58)a, which consists of the combination of *nanadaka* "somehow" with *yoo,* is most likely uttered on the basis of the visual appearance of the speaker's younger sister, while sentence (58)b, which consists of the combination of *nandaka* "somehow" and *yoo-na ki ga suru,* "I feel like," is solely based on the speaker's intuitive perception. The use of *yoo* is perfectly appropriate for both sentences, indicating that *yoo* is used when a speaker makes a judgment based on the visual appearance of a situation and/or his/her intuitive feeling. *Rashi* in sentences (58)d and (58)d, on the other hand, is not considered appropriate. *Nandaka* "somehow" is used based on the speaker's feeling and is not quite compatible with *rashii,* since it conveys information that the speaker has obtained from an outside source. When *ki ga suru* accompanies *nandaka, rashii* becomes totally unacceptable, demonstrating that the main function of *rashii* is to convey second-hand information, as with hearsay *soo* and is incompatible with a speaker's intuitive judgment alone. Thus, *yoo* involves a greater degree of speculation on the part of the speaker, that is, the degree of epistemic possibility expressed by *yoo* is greater than expressed by *rashi.*

If not only intuitive perception but also visual appearance is among the keys to using *yoo* (as was the case in the examination of *rashii* in the previous section), further delineation is necessary to distinguish the use of *yoo* and visual *soo.* As discussed above, visual *soo* is used when a speaker is exposed to a present situation and makes a judgment based on appearances. However, there are cases where a proposition based upon visual impression is incompatible with *yoo.* The following sentences exemplify such cases:

(59) a. Mite! Ki ga taore-soo-da.
look tree NOM fall down-AUX-COP
"Look! The tree is about to fall."

* b. Mite! Ki ga taoreru yoo-da.
look tree NOM fall down AUX-COP
"Look! It looks like the tree will fall down."

Yoo is appropriate in a situation that has already occurred or been altered into a different state, as was exhibited in sentences (55) and (56) in the previous section. This appropriateness is attributable to the fact that the speaker is able to witness the situation externally to make a judgment. Nevertheless, when an event is about to occur or change into a different situation at the present moment (i.e., the moment of the utterance), this is no longer the case. Visual *soo* is the only appropriate evidential. When a change is about to occur in the midst of an event, the immediacy of the event does not allow the speaker to process the whole event and make a judgment. This use of *soo* is infinitely close to the event itself and is, therefore, often replaced by the simple present form (in this case: *Ki ga taoreru!* "The tree will fall"). On the other hand, when the emphasis in a situation is not on change, that is, the predicate is a stative predicate, *yoo* as well as *soo* may appear:

(60) a. Kare wa samu-soo-da.
he TOP cold-AUX-COP
"He looks cold."

b. Kare wa samui yoo-da.
he TOP cold AUX-COP
"He looks cold."

The difference between sentence (60)a and (60)b may be difficult to detect merely from examining these sentences alone. Only through consideration of various circumstantial elements can subtle differences in meanings contained in these sentences be distinguished. For example, while *samu-soo-da* may be uttered as an immediate response to a present situation, *samui yoo-da* may be supported by a variety of visual and sensory inputs. The speaker may, for example, utter sentence (60)b after hanging up the phone as a way of telling a third party how the speaker felt about the interlocutor on the phone, or what the speaker heard from that person. It is not always necessary that a present situation occur before the speaker's eyes in order to use *yoo*. As discussed previously, *yoo* is inappropriate if the situation is occurring right in front of the speaker and the event is clearly describable. Here, *yoo* involves a much larger degree of the speaker's speculation than visual *soo*.

Just as present tense *yoo* is used to express a speaker's supposition based on visual evidence and sensory impressions, past tense *yoo* is also used to express a speaker's supposition, but one that was made sometime in the past. Please note the following example:

(61) Kinoo uchi e itte-mi-ta kedo daremo i-nakat-ta yo.
yesterday home LOC go-try-PST but nobody exist-NEG-PST VOC
"I went to (Miwa's) home yesterday, but no one was there."

Miwa san jikka e kaet-ta yoo-dat-ta yo.
Miwa parents' home LOC return-PST AUX-COP-PST VOC
"Miwa seemed to have gone back to her parents' house."

If the situation in example (61) were in the present, the sentence would be *Miwa wa jikka e kaet-ta yoo-da* "Miwa seems to have gone back to her parents' home." From the viewpoint of objective truth, however, it is not necessarily true that Miwa has gone back to her parents' home. It may be true, but it may not be true. The past tense *yoo* creates the same implication: Miwa may or may not have gone back to her parents' home. Both present and past tense forms of *yoo* stress the speaker's uncertainty of the proposition being true. This interpretation is very similar to that of the suppositional *kamoshire-nai* "may." In fact, *yoo* and *kamoshire-nai* are in many cases interchangeable. One could say, *Ame ga furu kamoshire-nai*, "It may rain," instead of *Ame ga furu yoo-da*, "It looks like it's going to rain" and convey roughly similar degrees of possibility. Some slight differences are that *yoo* is much more tightly bound to the current, visible, and sensible status of an event, and it arguably carries a stronger sense of the likelihood of the occurrence of the event. The following examples compare *kamoshire-nai* and *yoo:*

(62) a. Yoosuke wa hontoo no koto o shir-anai kamoshire-nai.
Yosuke TOP true GEN thing ACC know-NEG AUX
"Yosuke may not know the truth."

b. Yoosuke wa hontoo no koto o shir-anai yoo-da.
Yosuke TOP true GEN thing ACC know-NEG AUX-COP
"It looks like Yosuke does not know the truth."

The difference in these sentences is that *yoo* gives an impression that the judgment was made including Yosuke's appearance, whereas *kamoshire-nai* may not necessarily be visually based, but founded wholly upon speaker intuition. Hence *yoo* is located in between *kamoshire-nai* and visual *soo* in terms of the possibility of the proposition's truth and the speaker's degree of conviction as to its truth.

3.3.5 Summary of Evidentials

It has been shown that evidentials locate a speaker's remarks on a continuum between statements based on the speaker's supposition and statements based on outside information. Within this framework, evidentials have traditionally been examined in terms of subjective and objective judgment. For example, Teramura (1984) and Johnson (1990) concluded that visual *soo* indicates only a speaker's subjective judgment while hearsay *soo* does not involve the speaker's judgment, but presents information from the outside objectively. Thus, the two *soo* express totally opposite notions, with *yoo* and *rashii* located somewhat in between.

The question arises, however, as to how one can determine whether that which one hears is more objective than what one sees. As the proverb "seeing is believing" suggests, evidence based on visual appearance can be much more convincing and reliable than that based on other sensory information. When visual *soo* is said to express subjective judgment, it is subjective to another party but not to the speaker him/herself. Therefore, the dichotomy of subjectivity and objectivity does not seem to account precisely for the concept of modality (which is a speaker-oriented notion). The difference between these two, then, should be viewed in terms of the degree to which a proposition is true or realized, as viewed by the speaker.

The examination also shows that the manner in which a speaker views the situation—externally or internally—is related to a temporal factor and is a key to differentiating the selection of evidentials at the time of speech. *Rashii* tends to be chosen when a speaker has processed the situation in order to make a judgment. This process involves a great degree of sensory information, and in this sense, *rashii* is very similar to hearsay *soo* in that it is basically used to mark information a speaker obtained from an outside source. The slight difference, however, is that *rashii* is not an undiluted statement from a third party, but rather the speaker's supposition is, to a small extent, entwined in the conveyed information. Therefore, *rashii* is inappropriate for a situation currently progressing or a situation that requires an immediate judgment. *Yoo,* on the other hand, involves a great degree of visual impression of a situation, but the speaker cannot determine the truth of the proposition that describes the situation solely through visual observation. *Yoo* can involve both information from the outside and the speaker's supposition based on visual and other sensory impressions. The supposition expressed by *yoo,* however, is mostly derived from the appearance of a situation.

Visual *soo* is used to indicate a strong visual and sensory impression, and usually the object of the judgment is in the speaker's visual field. Even if the object is not visible it is always perceptually close to the speaker at the time of speech. This is supported by the fact that *soo* can be used to express that something is about to occur. *Soo* in this case is accompanied by a nonstative, change predicate. The speaker does not have time to process overtly the current information, and therefore that which is expressed is infinitely close to a statement that describes the truth. On the other hand, when *soo* co-occurs with a stative predicate, the speaker's degree of supposition of truth increases while visual and sensory perceptions provide strong, but not conclusive, evidence that the proposition is true.

Clearly, both visual and hearsay *soo* are related to epistemic necessity. This merely suggests, for example, that when someone says "The sea urchin looks delicious," it must appear to be delicious to the speaker, even if others might not believe that the sea urchin is edible. The appearance of the sea urchin appeals to the speaker to the extent that s/he strongly believes the proposition to be true.

Rashii expresses notions of both epistemic necessity and epistemic possibility, although epistemic necessity is the more important of the two. This is due to the fact that *rashii* is used when a speaker has time to process and judge the proposition's truth externally, combining his/her own sensation and information from outside. *Yoo* is used to express epistemic possibility, and the meaning is very close to that of *kamoshire-nai,* whose nonpast and past tense forms both express possibility. Evidentials in Japanese are thus concerned with outside information or a supposition based on visual and sensory impressions, or both. Nevertheless, they simultaneously express the possibility or the necessity of the proposition being true, and therefore express degrees of modality.

In order to summarize the functions of suppositionals and evidentials in terms of modality, and to prove the appropriateness of the description given for each auxiliary, tests are given from the syntactic and semantic viewpoints in the following section and confirm the validity of the discussion put forth thus far. The modals are examined in terms of: 1) compatibility with phrases that are used to express various degrees of speaker conviction; 2) compatibility with the actual situation described in the second half of the sentence that is counter to the situation in the first half of the sentence; and 3) compatibility with another modal.

3.4 COMPATIBILITY TESTS

3.4.1 Modals with Adverbs

This section scrutinizes the compatibility of modals with other phrases that express the degree of speaker conviction regarding the truth of a proposition, or that present the source of speaker judgment. These two types are distinguished by features that were made evident in the examination of suppositionals and evidentials. Adverbial phrases selected to test compatibility with modals expressing the former type are: *zettai*, "definitely;" *tabun*, "probably;" *osoraku*, "perhaps;" *moshikashi-tara*, "may be;" and *hyottoshi-tara*, "by chance."[30]

The phrases selected to test compatibility with a modal expressing the source of speaker judgment are used in combination with both suppositional and evidential modals. They are: *abunai*, "watch out!;" *ikkenshi-ta tokoro*, "apparently;" *watashi no shiru kagiri dewa*, "as far as I know;" and *uwasa ni yor-eba*, "from what I have heard." *Abunai!* is used to detect whether or not the modal is compatible with a situation that is about to be realized, that is, extremely close to an actual situation where speaker inference is unnecessary. *Ikkenshi-ta tokoro*, on the other hand, might give a speaker some room to speculate on the realization of an event based on a visual impression, that is, the outcome of the situation does not have to be fixed. *Watashi ga shiru kagiri dewa* indicates that the speaker believes that the proposition is true, but the source of the knowledge is unclear. S/he may have heard, read, and/or witnessed an event that supports the claim prior to uttering the proposition. *Uwasa ni yor-eba* indicates that the information is second-hand with a possibility of blending speaker inference, since this phrase implies a significant time expanse that allows the speaker to speculate on the truth of the proposition.

In addition, the phrases *nandaka* "I have a feeling that ~/somehow" and *dooyara* "something tells me/seems likely" are examined

30. These are used only in combination with suppositionals due to the difficulty in divorcing them from the propositional content. For example, most native speakers of Japanese interpret the sentence *Tabun ame ga furu soo-da* as "It will probably rain, I hear," where the adverb *tabun*, "probably," is included in the proposition, and they do not recognize the possibility of the interpretation that "I probably hear that it is going to rain." The survey did not reflect the use of these adverbs employed separately from the proposition; in other words, these phrases cannot be tested appropriately in combination with evidentials, since they are assumed to be in the domain of propositional content.

in combination with both types of modals. Although both phrases convey an appreciable degree of speaker uncertainty, this arises from a combination of the fact that the source of the information is unconfirmed together with his/her feeling and intuition over the truth of the proposition. Between the two, *nandaka* likely expresses a greater degree of a speaker intuitive judgment, since *nandaka* is more closely based on his/her feelings, while *dooyara* likely involves a broader range of evidence.

All of the above phrases were tested to discern whether or not they are compatible with the proposition *(Ano) ishi wa/ga, gake kara ochiru* "(That) stone falls/will fall off the cliff." The following table describes the result of the survey.

The following are representative examples for each type of compatible adverbial-modal sequence:

Suppositionals:

Ano ishi wa zettai gake kara ochiru hazu-da.

Ano ishi wa tabun gake kara ochiru ni chigai-nai.

Ano ishi wa osoraku gake kara ochiru daroo.

Ano ishi wa moshikashi-tara gake kara ochiru kamoshire-nai.

Ano ishi wa hyttoshi-tara gake kara ochiru kamoshire-nai.

Evidentials:

Abunai! Ishi ga gake kara ochi-soo-da.

Ikkenshita tokoro, ishi ga gake kara ochiru yoo-da.

Watashi ga shiru kagiri-dewa, ano ishi wa gake kara ochiru rashii.

Uwasani yore-ba, anoishi wa gake kara ochiru rashii.

Nandaka ano ishi wa gake kara oshi-soo-da.

Dooyara ano ishi wa gake kara oshiru rashii.

\+ indicates that more than 90% of respondents believe that the given adverb or phrase is compatible with the modal.

? indicates that the majority of respondents believe that the given adverb or phrase is compatible, but some respondents believe it to be incompatible.

?? indicates that some respondents believe that the given adverb or phrase is compatible, but the majority believe it to be incompatible.

\- indicates that more than 90% of respondents believe that the given adverb or phrase is incompatible.

(63)

Suppositionals	***Hazu***	***Nichigai-nai***	***Daroo***	***Kamoshire-nai***
Zettai	+	+	+	-
Tabun	??	+	+	??
Osoraku	-	?	+	?
Moshikashi-tara	-	-	?	+
Hyottoshi-tara	-	-	-	+
Abunai!	-	-	-	-
Ikkenshi-ta tokoro	-	+	-	??
Watashi ga shiru kagiri-dewa	+	?	+	??
Uwasa ni yor-eba	+	-	-	-
Nandaka	-	-	-	-
Dooyara	-	-	-	-
Evidentials	***(Ochi)soo***	***Yoo***	***Rashii***	***(Ochiru)soo***
Abunai!	+	-	-	-
Ikkenshi-ta tokoro	+	+	-	-
Watashi ga shiru kagiri-dewa	??	+	+	+
Uwasa ni yor-eba	-	?	+	+
Nandaka	+	+	??	-
Dooyara	??	+	+	??

On the basis of the compatibility test of suppositionals with adverbs, we observe that *hazu* is used to express the highest degree of speaker conviction and *kamoshire-nai* the lowest (as native speakers of Japanese could easily predict). *Ni chigai-nai* and *daroo* present very similar degrees of conviction in that both can co-occur with *zettai* "definitely" and *tabun* "probably," and as such express a high degree of speaker conviction. One decisive difference between the two, however, is the ability of *daroo* to co-occur with *osoraku* "perhaps" and *moshikashi-tara* "maybe"—two forms that express a low degree of speaker conviction and thus locate *daroo* at a lower degree of realization. The phrases in the second group, however, can variously co-occur with suppositionals. One notable characteristic is that *hazu* is compatible with *watashi ga shiru kagiri-de wa* "as far as I know" and *uwasa-ni yore-ba* "from what I have heard," but is not compatible with phrases that are based on a visual impression such as *abunai* "watch out!" and *ikkenshi-ta tokoro* "apparently," while *ni chigai-nai* is compatible with *ikkenshi-ta tokoro.* This result is consistent with the previous observation that *hazu* is inappropriate when the present condition has not yet become an objectively perceptive result. As for the last group, none of the suppositionals can co-occur with phrases that are used on the basis of a speaker's intuitive feelings.

The test results for suppositionals reveal that the notion of "supposition" is concerned with a speaker's knowledge that has already been processed in his/her mind, but not with a visible situation that can be instantaneously judged as to whether or not the proposition is true. Based on the behavior of suppositionals, the highest degree of speaker conviction among suppositionals is associated with outside information, rather than the intuitive judgment of the speaker. This statement is reasonable in that the truth of a proposition is more strongly related to evidence than intuition.

Among evidentials, it is relatively straightforward to discern that hearsay *soo* and visual *soo* seem to have little in common in terms of the meaning they express, but at root both of them convey that the speaker believes the proposition to be true. The degree of speaker conviction in the truth of a proposition suggested by hearsay *soo* is similar to that of *hazu,* since both can co-occur *with watashi ga shiru kagiri-de wa* and *uwasa ni yor-eba,* except that *hazu* involves speaker supposition whereas *soo* does not. Visual *soo* is the only modal that can co-occur with *abunai* and *ikken shi-ta tokoro,* each of which is based on visual impression. When visual *soo* co-occurs with *abunai,* however, the interpretation of the sentence is no longer "It looks like the stone falls off the cliff," but is rather "The stone is about to fall off the cliff," which is a proposition. In such a case, the speaker is referring to a situation undergoing change. As was discussed above, there is a clear difference in the interpretation of visual *soo* sentences, depending on the type of predicate with which it is associated. If visual *soo* is accompanied by a nonstative predicate (especially with a heterogeneous predicate, as discussed in the next chapter), then the sentence can be considered a proposition.

The function of *rashii* is similar to that of hearsay *soo,* since it is based on an auditory source. As for *yoo,* even though one of the keys to its use is primarily the appearance of a situation, *yoo* cannot be used when an event is in the process of occurring. For instance, it cannot be used when the situation is changing "now," as seen in the incompatibility of *yoo* with the phrase *abunai* "watch out!" *Yoo,* however, can co-occur with the phrase *ikken shi-ta tokoro* "apparently." Since this phrase gives a speaker some room to speculate on the realization of an event, it is perfectly compatible with *yoo.* This observation supports the idea that a speaker requires time to form an assumption in order to use *yoo.* Although the phrase *uwasa-ni yor-eba* "from what I have heard" also implies a time expanse between the time the speaker obtained the information and the time of speech, involvement of the direct auditory source makes the use of *yoo* with this phrase awkward. This explanation suggests that *yoo* is employed based on a speaker's visual impression when there is a certain degree of speculation regarding the situation, and its use creates a spatiotemporal distance between the speaker and the situation. Although the suppositionals and evidentials do not behave in exactly the same fashion (since suppositionals are concerned with a speaker's conviction whereas evidentials are concerned with a visual or auditory sense), both involve the notions of necessity and possibility in that they convey the degree to which the speaker believes in the realization of an event, based to some extent on a given source.

3.4.2 Modals with Actual Situations

In this section, the compatibility of both suppositional and evidential modals with an opposite situation described in the proposition is examined. This test was motivated by the discussion posed by White (1979) regarding the meaning of "could have," "may have," and "might have." The negative form of the verbs *furu*, "to fall," and *kau*, "to buy," are tested for their compatibility in counter situations that accompany a modal. The negative predicate appearing in the second half of each sentence is used in the sense of "In reality, it does/did not happen in the way it is/was supposed to happen." Therefore, only sentences expressing that the situation is/was about to be or supposed to be in the domain of the actual world are compatible with the second half of the sentence. Karttunen (1972) examined the same type of sentence combination and claimed that it is not acceptable in natural language to state in one sentence that something both is the case and possibly is not the case. This means that modals, which are concerned with the concepts of possibility and necessity, should not co-occur with a negative predicate in the second half of the sentence. It should also be noted that when the first half of the sentence represents an actual event, it results in a contradictory statement as a whole - for example, "I will go, but I will not go," or "I went, but I did not go." Please note the following examples:

Nonpast context

Suppositionals:

(64) ?? a. Kuruma o kau hazu-da keredo kaw-anai.
car ACC buy AUX-COP but buy-NEG
"I'm sure that I will buy a car, but I will not (buy a car)/I'm supposed to buy a car, but I will not (buy a car)."

* b. Kuruma o kau ni chigai-nai keredo kaw-anai.
car ACC buy AUX but buy-NEG
"I believe that I will buy a car, but I will not (buy a car)."

* c. Kuruma o kau daroo keredo kaw-anai.
car ACC buy AUX but buy-NEG
"I will probably buy a car, but I will not (buy a car)."

* d. Kuruma o kau kamoshire-nai keredo kaw-anai.
car ACC buy AUX-COP but buy-NEG
"I may buy a car, but I will not (buy a car)."

Evidentials:

(64) e. Ame ga furi-soo-da keredo, fur-anai.
rain NOM fall-AUX-COP but, fall-NEG
"It looks like it is about to rain, but it will not (rain)."

f. Ame ga furu soo-da keredo, fur-anai.
rain NOM fall AUX-COP but, fall-NEG
"I hear that it is going to rain, but it will not (rain)."

g. Ame ga furu yoo-da keredo, fur-anai.
rain NOM fall AUX-COP but, fall-NEG
"It looks like it is going to rain, but it will not (rain)."

?? h. Ame ga furu rashii keredo, fur-anai.
rain NOM fall AUX but, fall-NEG
"It seems like it is going to rain, but it will not (rain)."

Past Context

Suppositionals:

(65) * a. Kuruma o kat-ta hazu-da keredo kaw-anakat-ta.
car ACC buy-PST AUX-COP but buy-NEG-PST
"I am sure that I bought a car / I am supposed to have bought a car, but I did not (buy a car)."

* b. Kuruma o kat-ta ni chigai-nai keredo kaw-anakat-ta.
car ACC buy-PST AUX but buy-NEG
"I believe that I bought a car / I believe that I would have bought a car, but I did not (buy a car)."

* c. Kuruma o kat-ta daroo keredo kaw-anakat-ta.
car ACC buy-PST AUX but buy-NEG
"I probably bought a car / I would probably have bought a car, but I did not (buy a car)."

* d. Kuruma o kat-ta kamoshire-nai keredo kaw-anakat-ta.
car ACC buy-PST AUX-COP but buy-NEG
"It may be the case that I bought a car / I might have bought a car, but I did not (buy a car)."

Evidentials:

(65)	e.	Ame	ga	furi-soo-dat-ta	keredo	fur-anakat-ta.
		rain	NOM	fall-AUX-COP-PST	but	fall-NEG-PST

"It looked like it was about to rain, but it did not (rain)."

* f.	Ame	ga	fut-ta	soo-da	keredo	fur-anakat-ta.
	rain	NOM	fall-PST	AUX-COP	but	fall-NEG-PST

"I heard that it rained, but it did not rain."

g.	Ame	ga	fut-ta	yoo-da	keredo	fur-anakat-ta.
	rain	NOM	fall-PST	AUX-COP	but	fall-NEG-PST

"It looks like it rained, but it did not (rain)."

* h.	Ame	ga	fut-ta	rashii	keredo	fur-anakat-ta.
	rain	NOM	fall-PST	AUX/AUX	but	fall-NEG-PST

"It seems like it rained, but it did not (rain)."

The test results evidence an intriguing phenomenon regarding the interpretation of suppositionals with a proposition in a past context. The second English equivalent for sentences (65)a–(65)d, for example, allows one to envision an implied hypothetical situation even though a conditional clause such as "if I had money" is absent. In Japanese, however, these sentences alone are usually perceived as expressing the first English equivalent and are considered somewhat awkward. This is the reason that all of the suppositionals are considered incompatible with a past event in the second half of the sentence. In order to perceive these statements as logical, a supplementary conditional clause is necessary. For example, *(Okane ga at-tara) kuruma o kat-ta hazu-da ga, (nakat-ta kara), kaw-anakat-ta* "(If I had money), I was supposed to have bought a car, but (since I did not have money), I did not buy one" is a sensible sentence.[31] (Supplied information is indicated in the parentheses.) The same thing can be said for *ni chi-*

31. In this statement, *hazu* functions as a modal. However, when it partakes in the past tense form, *kau hazu-dat-ta*, "I was supposed to buy," it no longer expresses a counterfactual meaning. This observation supports the examination of *hazu* in chart (11) in section 3.2.1, in which *hazu-dat-ta* does not function as a modal.

gai-nai, daroo, and *kamoshire-nai.* Regarding the compatibility of the two sentences, White (1979) employed logical ambulate to solve this problem. Taking the example of "may," he demonstrated that "X may have happened" and "X may not have happened" are compatible; therefore, "X may have happened" cannot imply "X did happen," but is compatible with its falsity. This means that "I might have bought" is not compatible with "I bought," but with "I did not buy." All suppositionals in past context can help to express a condition that is not in fact met; therefore, sentences (65)a–(65)d are sensible sentences when sufficient context is provided. *Kuruma o kat-ta ni chigai-nai* can be interpreted as either "I believe that I bought a car (I must have bought a car)" or "I believe that I would have bought a car," with the former case incompatible with *kaw-anakat-ta* "I did not buy." Similarly in English, "must have" does not express a counterfactual situation.

Coates (1983) examined the epistemic "must" and perfective aspect, noting that "while epistemic 'must' can refer to states and activities in the past, present and future, the modal predication is unaffected: I'm sure that X was~ (past); I'm sure that X is ~ (present); and I'm sure that X will be ~ (future)" (p. 45). This observation demonstrates that "I must have bought a car" is equivalent to "I'm sure that I bought a car," but not to "I'm sure that I would have bought a car" in English. As mentioned above, however, the meaning of a predicate in the past tense form in Japanese, can be ambiguous, expressing either "I bought" or "I would have bought" for *kat-ta,* for example, when used with a subordinate clause. This problem can be avoided if the aspectual form *te-iru* is used with the verb, such as *Kuruma o katte-i-ta daroo* "I would have bought a car." This means that in order to clearly express a counterfactual situation, a stative predicate that creates a spatiotemporal distance between the situation and the speaker is required. A detailed discussion in this regard is found in chapter 5.

The above observation demonstrates that suppositionals co-occurring with a proposition in past context are not merely expressing a speaker's supposition toward a past event; they also help in understanding a counterfactual (hypothetical) meaning implied in the sentence. The degree to which the event might have been realized depends on the nature of suppositionals. Thus, we catch a glimpse of the fact that the notions of modality and hypotheticality are deeply related.

Evidentials with a proposition in past context exhibit a different aspect in their behavior compared with suppositionals. An interesting phenomenon concerns the use of *soo,* resulting in completely opposite results for its two meanings as indicated in sentences (65)e and (65)f. *Ame ga fut-ta soo-da* "I hear that it rained" is perceived as an actual event conveyed to the speaker for him/her to believe, and therefore *fur-anakat-ta* invites contradiction. This observation indicates that hearsay *soo* is perceived as a constituent of the proposition, which is not concerned with the speaker's conviction as to its truth; it is a statement that the speaker believes to have been actualized. Therefore, it can be said that hearsay *soo* is closer to the actual world than visual *soo,* that is, it expresses a slightly greater degree of certainty. Due to the extreme proximity (i.e., infinite closeness without quite reaching the point of realization) between visual *soo* and the actual situation described in the propositional content, visual *soo* can invite a counter situation in the second half of the sentence. Neither *soo* is used to express a hypothetical situation.

A difference is also seen in the manner that *yoo* and *rashii* are interpreted; *yoo* is compatible with *fur-anakat-ta* "did not rain," but *rashii* is not. The involvement of visual impression in *yoo* may make the situation more conceivable, while the involvement of hearsay and reliance on outside information creates contradiction with *fur-anakat-ta* "did not rain." This suggests that the function of *rashii* is close to that of hearsay *soo.*

Regarding modals with a proposition in nonpast context, only *hazu* is somewhat ambiguous in its interpretation. The rest of the suppositionals are considered incompatible with the second half of the sentence. *Hazu* in sentence (64)a can be interpreted as *Kau-kotoni-natte-iru* "I am supposed to buy" in this context, and in such a case, *hazu* acts as a constituent of the proposition and is therefore nonmodal (as discussed in section 3.2.1). This ambiguity is clearly expressed in the uncertainty of the survey results. *Kotoni-natte-iru* is compatible with the speaker's will against the preplanned activity (will not buy a car), but *hazu* as a modal expresses a strong belief regarding the purchase of the car and contradicts the idea of not purchasing a car. *Ni chigai-nai, daro* and *kamoshire-nai* create contradiction with the second half of the sentence. The speaker's belief/judgment, regardless of its degree, once made toward the proposition does not co-occur with *with* the speaker's will against the proposition.

Evidentials, however, are all compatible with the second half of the sentence. This result points to the difference in the nature of evidentials and suppositionals in that the notion of evidentials is equal to the notion of externals, while the notion of suppositionals is equal to the notion of internals. The judgment made by the speaker does not originate from the speaker for evidentials, but does for suppositionals. The notion of evidentials, therefore, precisely corresponds to Kamio's theory of territory of information, where his theory is basically equivalent to the theory of evidentiality. When the source of information is outside the speaker's domain, even if reality is counter to such information, the speaker can make a statement combining two opposite situations. For, example, *Ame ga furu rashii ga fur-anai* means that "Although I hear that it is going to rain, but against what I hear, the reality is not the case." The first half of the sentence does not genuinely originate from the speaker in the same manner as a clause accompanying the suppositionals *ni chigai-nai, daroo,* and, *kamoshire-nai.* This is why *hazu,* which also indicates that the speaker obtained the information from an outside source, behaves somewhat similarly to evidentials.

3.4.3 Modals with Another Modal

The compatibility of both suppositional and evidential modals with another modal is examined to explain their syntactic behavior. Since modals are all considered to be in the domain of modal content, syntactically they are not supposed to invite another modal since such addition would force the existing modal into the propositional content. Note the following examples:

Nonpast context

Please note that the English equivalents do not necessarily reflect the appropriateness or inappropriateness of the Japanese sentences.

Suppositionals:

(66) ? a.	Taroo	wa	Hanako	ni	au	hazu	daroo.
	Taro	TOP	Hanako	DAT	meet	AUX	AUX

"It is probably the case that Taro must meet Hanako."

* b.	Taroo	wa	Hanako	ni	au	ni chigai-nai		kamoshire-nai.
	Taro	TOP	Hanako	DAT	meet	AUX		AUX

"It may be the case that Taro must meet Hanako."

* c.	Taroo	wa	Hanako	ni	au	daroo	ni chigai-nai.
	Taro	TOP	Hanako	DAT	meet	AUX	AUX

"I believe that Taro will probably meet Hanako."

* d.	Taroo	wa	Hanako	ni	au	kamoshire-nai	hazu-da.
	Taro	TOP	Hanako	DAT	meet	AUX	AUX

"It must be that Taro may meet Hanako."

Evidentials:

(A situation occurring in front of the speaker.)

(66) * e.	Ame	ga	furi-soo	kamoshire-nai.
	rain	NOM	fall-AUX	AUX

"It may be the case that it looks like it is about to rain."

* f.	Ame	ga	furu	soo	daroo.
	rain	NOM	fall	AUX	AUX

"It may be the case that it looks like it is about to rain."

* g.	Ame	ga	furu	yoo	ni chigai-nai.
	rain	NOM	fall	AUX	AUX

"I'm pretty sure that it looks like it is going to rain."

* h.	Ame	ga	furu	rashii	yoo-da.
	rain	NOM	fall	AUX	AUX-COP

"I'm pretty sure that it seems like it will rain."

Modals in past tense forms with another modal

Note that sentence "c" does not have an English equivalent, and the English equivalents for "f," "g," and "h" do not reflect the inappropriateness of the original Japanese sentence.

Suppositionals:

(67) a. Taroo wa Hanako ni au hazu-dat-ta kamoshire-nai.
Taro TOP Hanako DAT meet AUX-COP-PST AUX
"It may be the case that Taro was supposed to meet Hanako."

?? b. Taroo wa Hanako ni au ni chigai-nakat-ta hazu-da.
Taro TOP Hanako DAT meet AUX-PST AUX-COP
"It must be that Taro must have met Hanako."

* c. Taroo wa Hanako ni at-ta daroo ni chigai-nai.
Note that the modal daroo does not partake in past tense form.

* d. Taroo wa Hanako ni at-ta kamoshire-nakat-ta daroo.
The suppositional *kamoshire-nai* does not follow the nonpast tense form of a predicate.

Evidentials:

(67) e. Ame ga furi-soo-dat-ta ni chigai-nai.
rain NOM fall-AUX-COP-PST AUX
"It must have been the case that it looked like it was about to rain."

* f. Ame ga furu soo-dat-ta daroo.
rain NOM fall AUX-COP-PST AUX
Hearsay *soo* does not partake in the past tense form.

* g. Ame ga furu yoo-dat-ta kamoshire-nai.
rain NOM fall AUX-COP-PST AUX
"It may be the case that it looked like is going to rain."

* h. Ame ga furu rashikat-ta hazu-da.
rain NOM fall-PST AUX-PST AUX-COP
"I'm sure that it seemed like it is going to rain."

In nonpast context, none of the modals can co-occur, which verifies that when a speaker makes a judgment regarding a situation, it is not possible to express two different degrees of conviction, such as "maybe" and "must be" simultaneously, that is, the different degree of speaker conviction is irreconcilable as of the time of speech. None of the modals function as part of a proposition except *hazu,* which for some individuals can be interpreted the same as the proposition *-kotoni-natte-iru* "is supposed to."

When partaking in past tense forms, however, *hazu* certainly functions as part of the proposition and therefore can accompany another modal. This behavior is common to the past tense forms of *ni chigai-nai* and visual *soo,* although the appropriateness of the use of *ni chigai-nakat-ta* with *au* "to meet" was questionable to some extent.[32] Visual *soo* in sentence (67)e is part of a proposition that describes an event in past context and can invite a modal, implying that the speaker speculates the possibility of the truth of such a past event. Visual *soo* with a proposition in past context objectively describes a situation that was infinitely close to an actual situation and is part of the propositional content. Modals in past tense form no longer express speaker conviction over the realization of an event; they no longer function as modals, but rather are part of the proposition.

Another interesting observation regarding the use of hearsay *soo,* in both nonpast and past tense forms, is that it can follow another modal, as seen in (66)f and (67)f, but it cannot invite one. As discussed in section 3.3.1, when hearsay *soo* follows a modal—*hazu-da,* for example—it places the preceding modal in the domain of the proposition, expressing the conviction of a third party and not that of the speaker. The sentence *Saifu wa hikidashi ni aru hazu-da soo-da* "I hear that the wallet must be in the drawer" describes the location of the wallet as believed by a third person, and the speaker is relaying the conviction of the third party. Therefore, hearsay *soo* can follow another modal.[33] This is not the case, however, when a modal follows hearsay *soo.* The speaker's strong conviction (supposedly conveying a statement of

32. The interpretation of *au ni chigai-nakat-ta* is *au koto ni machigai wa nakat-ta,* "There was no mistaking the fact that they were to meet."

33. One exception is *daroo*. The sentence *Saifu wa hikidashi ni aru daroo soo-da,* "I hear that I think it is probably the case that the wallet is in the drawer" is considered an inappropriate sentence, since *daroo* cannot be viewed as the supposition of a third person. *Daroo* is used only in the speaker's domain.

fact) is incompatible with a modal that indicates speaker uncertainty; therefore, hearsay *soo* has the same feature as other modals in that it cannot invite another modal.

3.4.4 Summary of Test Results

Language use can vary depending on the environment in which speakers find themselves. Since epistemic modals are concerned with the degree of speaker conviction, and the degree may differ from person to person, the survey helps to define to what extent native speakers of Japanese perceive a given use of the language as appropriate or as inappropriate. Although the degree of speaker conviction expressed in the use of suppositionals may be ordered intuitively by native speakers of Japanese, the examination of the compatibility of modals with adverbs, counter situations, and other modals were supporting factors in such native speaker intuition, as well as the degree of speaker conviction of suppositionals in relation to that of evidentials. It was found that both types of modals involve different elements in determining the degree of necessity and possibility : suppositionals and evidentials undoubtedly function as modals with a proposition in nonpast context, expressing the notions of necessity and possibility that are concerned with possible/nonactual world situations. This fact is especially evident in the compatibility of modals with other modals. Only visual *soo* with a proposition in past context can invite another modal, where the modal *soo* itself has to occur in the past tense form.

Hazu and visual *soo* share similar behavior. Since *hazu* requires evidence that should be processed and externally observable from the speaker's viewpoint, the use of *hazu* is closer to that of evidentials than other suppositionals. The compatibility test of modals with a counter situation in (64) demonstrates the similarity of *hazu* with evidentials: while the suppositionals *ni chigai-nai, daroo,* and *kamoshire-nai* are incompatible with a counter situation, evidentials and *hazu* are compatible to a great extent. The use of these two modals indicates that the realization or the truth of an event is certain from the speaker's viewpoint and therefore they express a high degree of epistemic necessity. The difference is that the use of *hazu* is based on various sources available to the speaker and the evidence should already be externally observable at the time of speech, while use of visual *soo* is primarily based on perceptible appearance. Therefore, visual *soo* is related more closely than *hazu* to an actual situation described in a proposition and involves a greater degree of speaker conviction than *hazu*. Such a claim is evidently supported by the appropriateness of sentences (64)*e Ame ga furi-soo-da keredo, fur-anai* "It looks like it is about to rain, but it will not rain" and (65)*e Ame ga furi-soo-dat-ta keredo, fur-anakat-ta* "It looked like it was about to rain, but it did not rain," where visual *soo* is compatible with a counter situation in both nonpast and past context. As shown in the English equivalents, both sentences indicate that the situations have almost reached the point of change.

Regarding visual and hearsay *soo,* while both express a high degree of epistemic necessity, there are two sets within the test that indicate a clear difference between them: 1.) *soo* with a counter situation, and 2.) *soo* with another modal accompanied by a proposition in past context; see sentences (65)e, (65)f, (67)e, and (67)f. Visual *soo* in sentences (65)e, *Ame ga furi-soo-dat-ta keredo, fur-anakat-ta,* "It was about to rain, but it did not rain" and (67)e, *Ame ga furi-soo-dat-ta ni chigai-nai,* "It must have been the case that it looked like it was about to rain," occurs in the past tense form and is considered a constituent of the proposition. Hearsay *soo* in sentence (65)f *Ame ga fut-ta soo-da,* "I heard that it rained," on the other hand, is used to express that the statement is based on an outside source, and the speaker is conveying the information as truth in an indifferent manner. The reason it is not compatible with the counter situation, *Fur-anakat-ta,* "It did not rain," lies in the actuality of the statement, that is, the statement can be considered almost identical to the declarative statement, *Ame ga fut-ta,* "It rained," which contradicts the opposite factual statement *Fur-anakat-ta,* "It did not rain." Hearsay *soo,* therefore, is closer to the domain of the proposition (actual world situation) than visual *soo.*

Ni chigai-nai and *rashii* were also compared in terms of the degree of speaker conviction. Although the test cannot be performed in exactly the same method for these modals due to the fact that their use is based on different information sources, in using *rashii,* a speaker's conviction is intertwined with the outside source to a greater degree than in the case of *ni chigai-nai.* The inability of *ni chigai-nai* and the ability of *rashii* to co-occur with a phrase that expresses a low degree of speaker conviction (e.g., *dooyara* "somehow") proves that *ni chigai-nai* involves a higher degree of speaker conviction than *rashii.* *Ni chigai-nai* is used when a speaker is convinced by various other sources including his/her intuitive judgment; therefore, the source of the information is perhaps richer than in the case of *rashii.*

Daroo, kamoshire-nai, and *yoo* are found to be pure modals, since none of them are compatible with counter situations or another modal in any context. Among these three modals, the ability of *daroo* to co-occur with phrases that express a high degree of speaker conviction, such as *zettai, kanarazu,* and so forth correspondingly allows *daroo* to express the highest degree of epistemic necessity. Furthermore, *daroo* is a purely speaker-oriented expression, since it cannot occur with hearsay *soo.* Between *yoo* and *kamoshire-nai,* the former is more visually oriented, whereas the latter is intuition-oriented, as observed in the compatibility of each with *ikkenshi-ta tokoro,* "apparently." Also, the test that examines whether or not the first half of the sentence containing a given modal is compatible with the second half containing the same modal [examples (60)] supports the notion that *kamoshire-nai* conveys the lowest degree of speaker conviction regarding the realization of an event.

The following chart summarizes the degree of modality expressed through epistemic modals.

(68)

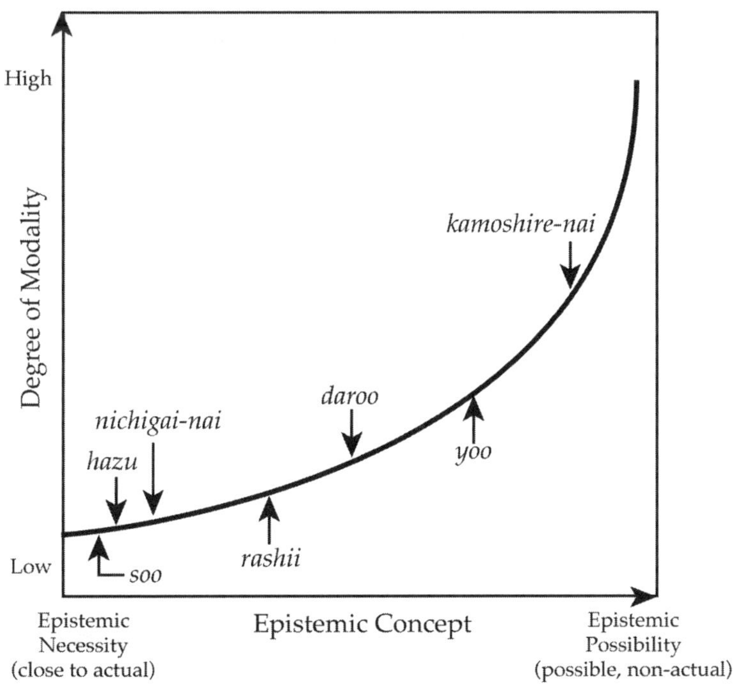

In figure (68), *soo* represents both visual and hearsay *soo*. If close attention is given to this point based on the examination conducted in this chapter, it is evident that the highest degree of epistemic necessity is expressed by hearsay *soo*, followed by visual *soo* with a nonstative predicate, and thirdly by visual *soo* with a stative predicate.

3.5 DEONTIC MODALITY

Deontic modality is concerned with a speaker's attitude, which in turn is related to the notions of permission, prohibition, and obligation. The meanings associated with deontic modality differ from those of epistemic modality, which is concerned with how a speaker's knowledge or beliefs are expressed in relation to the necessity or possibility of a proposition. Deontic modality is concerned with "the necessity or possibility of acts performed by morally responsible agents" (Lyons 1977, p. 823). The common point is that both epistemic and deontic modality involve notions of necessity and possibility.

Japanese deontic modality is quite distinct from epistemic modality in that the constructions that express deontic modality all appear in the propositional content. This syntactic characteristic is consistent with the theory of territory of information proposed by Kamio (1990) where he points out that the permissive form, for example, is a direct expression that falls in the domain of the speaker's territory. There are no modals that are equivalent to English deontic modals, that is, there is no deontic modal in Japanese whose syntactic behavior is consistent with epistemic modals. One might consider *beki* "should" a modal due to its independent syntactic and semantic features. *Beki,* however, falls within the domain of propositional content in both nonpast and past tense forms as it can invite an epistemic modal such as *Iku beki-dat-ta hazu-da* "I'm sure that I should have gone." Based on the premise proposed herein, Japanese deontic constructions are thus in the domain of the proposition, at least in their syntactic behavior, even though they are concerned with the notions of necessity and possibility. Permission is represented by the *-te-mo ii* construction, prohibition by *-tewa ik-e-nai,* and obligation by *-nake-reba-nar-anai.* These are provisionally termed "deontic constructions" hereafter. The following examples (69)–(71) represent these deontic constructions, and (72)–(74) demonstrate that such are in the domain of propositional content:

(69) Uchi e kaette-mo ii.
home LOC return-even if all right
"It's all right if you go home (you may go home)."

(70) Uso o tsuite-wa ik-e-nai.
lie ACC tell-COND go-POT-NEG
"It cannot go if you tell a lie (do not tell a lie)."

(71) Shukudai o das-anake-reba nar-anai.
homework ACC turn in-NEG-COND go-NEG
"It will not do if I do not turn in my homework (I must turn in my homework)."

(72) Konnani mazui-no-da-kara,
this much nasty-COMP-COP-because

zenbu tabe-nakute-mo ii hazu-da.
all eat-NEG even if all right AUX-COP
"Since the taste is so bad, it must be that we do not have to eat it all."

(73) Sakki shachoo ga suwatte-i-ta kara,
a little while ago president NOM sit-PROG-PST because

koko wa watashitachi ga suwatte-wa ik-e-nai ni chigai-nai.
here TOP we NOM sit-COND do-POT-NEG AUX
"Since the president was sitting here a little while ago, I'm pretty sure that we should not sit here."

(74) Rainen wa nihon e kaer-anake-reba-nar-anai kamoshire-nai.
next year TOP Japan LOC return-NEG-COND-do-NEG AUX
"I may have to go back to Japan."

The behavior of Japanese modals in sentences (72)–(74) shows that deontic modality is secondary to epistemic modality, since deontic constructions are all in the domain of the proposition and therefore can be followed by an epistemic modal. In the following examples, the degree of necessity and possibility expressed by deontic constructions is examined by testing whether or not the given form is compatible with an actual event. This is the same test as that proposed for epistemic modals. When a sentence that expresses deontic modality (the first sentence) is compatible with an objective statement spoken by the same speaker (the second sentence), the deontic construction creates a strong degree of necessity that the proposition will be actualized from the speaker's viewpoint. When, on the other hand, the first sentence is not compatible with an objective statement, it expresses a low degree of necessity that the proposition will be actualized. Please note the following:

(75) a. Te o age-ro. Samonaito, utsu zo.
hand ACC raise-IMP or shoot VOC
"Put up your hands, or I will shoot."

b. Kore o yom-anake-reba nar-anai.
this ACC read-NEG-COND do-POT-NEG
Samonaito, shiken ni ochiru.
or test DAT fail
"I must read this, or I will fail the exam."

c. Kono fugu wa tabete-wa ik-e-nai. Taberu-to, shinu.
this blow fish CONT eat-COND do-POT-NEG eat-COND die
"You must not eat this blowfish, or you will die."

* d. Kore o yonde-mo ii. Samonaito, shiken ni ochiru.
this ACC read-even if all right or test DAT fail
"It's all right if you read this, or you will fail the exam."

* e.	Kore	o	yom-anakute-mo	ii.
	this	ACC	read-NEG-even if	all right

Yomu-to	shiken	ni	ochiru.
read-COND	exam	DAT	fail.

"You do not have to read this. If you read it, you will fail the exam."

The deontic constructions in sentences (75)a, b, and c are all compatible with an objective statement. This compatibility demonstrates that the deontic constructions "obligation" and "prohibition" express a strong degree of necessity, while the "permission" constructions do not. In the next two sections a detailed comparison of these deontic constructions is given in relation to the English deontic modals "must" and "may."

3.5.1 "Must" in Japanese

Coates (1983) postulates that the English language "must" has two main meanings, a root meaning (obligation/necessity) and an epistemic meaning (logical necessity/confident inference). Root "must," despite its range of meanings, can be restated as "it is necessary that~," whereas epistemic "must" conveys the speaker's confidence in the truth of what s/he is saying, based on a logical process of deduction of facts known to him/her. Depending on the situation and circumstances, a Japanese deontic construction may sometimes be interpreted as "should" and at other times as "must." Coates (1983), however, asserts that "in the case of 'must' the speaker expects to be obeyed, but in the use of 'should' there is no such expectation. At its strongest 'should' takes on the meaning of moral obligation, or duty. At its weakest, it merely offers advice, if subjective, or describes correct procedure, if objective" (p. 59).

In Japanese, two modals and two constructions representing deontic modality—as well as variations of all four of these—roughly correspond to the single English modal "must." Although the English "must" exhibits two meanings, Japanese uses different modals and constructions to express these meanings. They are *hazu* and *ni chigai-nai* for epistemic modality, and *beki* and *-nakere-ba-nar-anai/-nakute-wa-ik-e-nai* for deontic modality. These can all express the modal content of "must." The objective in this section is to clarify the subtle differences among them. First note the following examples:

(76) a.

Konohen	ni	toire	ga	aru	hazu-da.
around here	LOC	bathroom	NOM	exist	AUX-COP

"I'm sure that there is a rest room around here./There should be a rest room around here."

b.

Konohen	ni	toire	ga	aru	ni chigai-nai.
around here	LOC	bathroom	NOM	exist	AUX

"I'm sure there is a bathroom around here./There must be a rest room around here."

c.

Konohen	ni	toire	ga	aru	beki-da.
around here	LOC	bathroom	NOM	exist	AUX-COP

"A bathroom should be around here."

d.

Konohen	ni	toire	ga	nake-reba-nar-anai.
around here	LOC	bathroom	NOM	exist-NEG-COND-do-NEG

"A bathroom must be here."

The first two modals fall under the category of epistemic modality, and the last two under the category of deontic modality. When *beki* is used, it expresses an obligation as strongly as *-nake-reba-nar-anai.* The difference between them is that *beki* implies some kind of prerequisite idea or information, while *-nake-reba-nar-anai* or *-nakute-wa ik-e-nai*[34] is based purely on the speaker's judgment. *Beki* means "duty," or "obligation" as it is related to what one is supposed to do in the society in which s/he lives. Therefore, it seems that when *beki* is

34. There are actually four expressions that are a combination of these two expressions. *Nake-reba-nar-anai, -nake-reba-ik-e-nai, -nakute-wa nar-anai, and -nakute-wa ik-e-nai.* These are all used to express the meaning "must." Details on the differences among these uses are not taken up in this book. However, there is a tendency to use *-nar-anai* when a speaker obliges an action on him/herself and to use *-ik-e-nai* when a speaker obliges someone else to act a certain way. This may be supported by the fact that the *-ik-e-nai* form is primarily used to express the meaning "you must not or you should not," as expressed in the following examples:

(a)

Koko	e	kite-wa	ik-e-nai.
here	LOC	come-COND	go-POT-NEG

"It won't do if you come here. Literally: You shouldn't come here."

used, the feeling is that a judgment is being made based on social expectations. The speaker is trying to convey that the basis for imposing the obligation on the interlocutor is not due to the speaker's own judgment, but to social expectation, such as in "you are supposed to do it this way in this culture." It is not an easy task to demonstrate a structural difference that clarifies this; nevertheless, the following expressions illustrate it relatively well:

(77) a.

Sanzan kara	osewa-ni nat-ta	sensee	no	owakarekai-da
thoroughly so	indebt-PST	teacher	GEN	farewell party-COP

kesseki suru	beki-ja-nai.
attend	AUX-COP-NEG

"Since it is a farewell party for the teacher to whom we are deeply indebted, you must/should attend."

b.

Sanzan kara	osewa-ni nat-ta	sensee	no	owakarekai-da
thoroughly so	indebt-PST	teacher	GEN	farewell party-COP

kesseki shite-wa	ik-e-nai
absent-COND	go-POT-NEG

"Since it is a farewell party for the teacher to whom we are deeply indebted, you must/should attend."

Sentence (77)a gives the impression that the speech action was based on morals and societal expectation; the interlocutor is expected to morally obey what s/he was told to do. In sentence (77)b the

---

(b)

Moo	kuji-da	kara	kaer-anakere-ba -nar-anai/-ik-e-nai.
already	9 o'clock-COP	because	return-NEG-COND-do-NEG

"It's already 9 o'clock, so I/you must go home."

Although it can be a matter of personal preference, generally speaking, a listener will understand that the speaker has to go home when s/he utters *kaer-anakere-ba-nar-anai*. On the other hand, the listener will understand that the speaker is telling someone to go home when *kaer-anakere-ba-ik-e-nai* is uttered. In this regard, Jorden (1990) explains that the difference is that *ik-e-nai* is slightly more forceful and direct than *nar-anai* (Part 3, p. 95).

speaker also obliges the interlocutor to carry out an action. The sentence, however, does not strongly imply that a societal expectation is involved, but rather that the speech act is the result of the speaker's own judgment. The following chart describes how the English language "must" is interpreted in Japanese:

(78)

	Epistemic Modality	**Deontic Modality**
↑ Necessity	*Hazu* Prerequisite evidence	*Beki* Social expectation
	Ni chigai-nai Strong intuitive assumption	*-nake-reba-nar-anai* Speaker's judgment of obligation

Both *hazu* and *ni chigai-nai* in epistemic modality and *beki* and *-nakere-ba-nar-anai* in deontic modality are in the domain of the notion of necessity, varying in the degree to which each expresses necessity. The difference between *beki* and *-nake-reba-nar-anai* may not imperatively be concerned with the different degree of necessity, since they are almost interchangeable and both express a speaker's sense of obligation. *Beki,* however, is used on the basis of social expectation, which is similar to *hazu* in that social expectation can be considered prerequisite evidence in the truth or realization of the events in question. In this sense, *beki* can be said to express a higher degree of necessity than *-nake-reba-nar-anai.*

3.5.2 "May" in Japanese

The English modal "may" expresses both an epistemic possibility and a deontic possibility. "May" is used to express two primary meanings: epistemic possibility, that is, a low degree of possibility of the proposition (nonactual world situations); and a deontic meaning, which is roughly equivalent to "it is permissible/allowed for X," or, peripherally, "it is possible for X to do ~" (Coates 1983). Japanese has different forms to express each meaning. They are *kamoshire-nai* and *-te-mo ii*. Please note the following examples:

(79) a. Anata wa koko ni ite-mo ii.
you TOP here LOC stay-even if OK
"You may stay here."

b. Konkai wa Sakai ga makeru kamoshire-nai.
this time TOP Sakai NOM lose AUX
"This time, iron chef Sakai may lose."

While the epistemic "may," *kamoshire-nai,* does not directly involve the interlocutor, the deontic "may," *-te-mo ii,* involves the second person directly. Just as the English "may" expresses the possibility of the proposition's realization by the interlocutor, the Japanese *-te-mo ii* expresses these same notions. Although the English "may" does not invite negation, *-te-mo ii* follows after a negative form, *-nakute-mo ii* "it's okay if you do not~." When *-nakute-mo ii* is used, however, it involves not only the notion of possibility, but also of necessity. Compare the affirmative and negative forms in the following sentences:

(80) a. Suki-nara motto tabete-mo ii.
like-COND more eat-even if OK
"If you like it, it's okay if you eat more."

b. Kirai-dat-tara tabe-nakute-mo ii.
dislike-COP-COND eat-NEG-even if OK
"If you do not like it, it's okay if you do not eat it."

Sentence (80)a expresses that the interlocutor is allowed to eat more. This sentence can be restated as "It is possible (for the interlocutor) to eat more." Similarly, *tabe-nakute-mo ii* in (80)b can be restated as "It is possible (for the interlocutor) not to eat," but it can also be restated as "It is not necessary (for the interlocutor) to eat." This communicates the general correlation that "It is possible that not-*p*" is functionally equivalent to "It is not necessary that *p*." (Refer to chapter 2, example [1], as well as Karttunen [1972].) Therefore, *nakute-mo- ii* is related to both necessity and possibility. Nevertheless, the former interpretation (necessity) is more appropriate for sentence (81)b. This is attributable to the fact that the notion of "possibility" also communicates a sense of "ability." In English, "you may eat" is interchangeable with "you can eat," but "you do not have to eat" is not interchangeable with "you do not have the ability to eat." Japanese has the same pattern. Please note the following examples:

(81)	a.	Tabete-mo ii	=	Tabe-rareru
		eat-even if-okay	=	eat-POT
		"It's okay if you eat."	=	"You can eat."
	b.	Tabe-nakute-mo-ii	≠	Tabe-rare-nai
		eat-NEG-even if-okay	≠	eat-POT-NEG
		"It's okay if you do not eat."	≠	"You are not able to eat./ You are not allowed to eat."

The potential form *~rareru* indicates not only ability, but also permission in these examples. When an interlocutor's desire (e.g., "not to eat") is allowed by the speaker, the notion of necessity (i.e., "It is not necessary for you to eat") is a suitable interpretation of the sentence. The following chart is a summary of the deontic forms:

(82)

Ie "Say it."	N	Necessity
Iu beki "I/you must/should say it."	N	↓
Iw-anake-reba-nar-anai "I/you must say it."	N	↓
Iw-anakute-mo ii "It's okay is you don't say it."	N/P	↓
Itte-mo ii "I/you must/should say it."	P	Possibility

3.5.3 Summary of Deontic Modality

In Japanese, deontic modality is formulated only by taking the semantic representation into consideration. The syntactic behavior of deontic modality is entirely different from that of the English language, and the behavior as constituents of a proposition clearly demonstrates that deontic modality is secondary to epistemic modality. Deontic constructions are Kamio's "bare-form expressions" and do not function as Fauconnier's "space-builders." They express not-yet-realized world situations, yet it is the interlocutor who is forced or is given permission to realize the event by the speaker; it is not the case that the speaker is speculating the realization/truth of the event. Deontic constructions thus involve the speaker's direct participation in the event's realization.

Examination of the representative deontic constructions revealed that such are predominantly concerned with the notion of necessity, which is close to an actual situation from a speaker's viewpoint. The degree of necessity expressed by *beki* "should" and *-nake-reba-nar-anai* "must" somewhat parallels that of the epistemic modals *hazu* and *ni chigai-nai* in that both *beki* and *hazu* are used based on prerequisite evidence in the truth or realization of events, while *-nake-reba-nar-anai* and *ni chigai-nai* are used based on the speaker's own deduction.

The notion of possibility is also expressed by the deontic construction *-te-mo ii,* which involves only the notion of possibility that corresponds to the use of *kamoshire-nai.* In fact, the degree of possibility of an event's realization for *-te-mo ii* is unknown. The interlocutor may or may not actualize the event; it remains to be seen by the speaker, since whether or not the interlocutor actualizes the event is controlled by the interlocutor, not by the speaker. This notion of "uncontrollability by the speaker" is thus the point which both epistemic modality and deontic modality commonly share. Modality refers to a world in which the speaker cannot completely manipulate the truth/realization of the event but can only observe events as a third person—a world in which a spatiotemporal distance is created between the speaker and the event in question; therefore, modality refers to not-yet-realized situations, that is, possible/nonactualized world situations.

3.6 SUMMARY: REFOCUSING THE DEFINITION OF MODALITY IN JAPANESE

Epistemic modality involves speaker-oriented notions of possibility and necessity that are not expressed through deontic modality. Deontic modality is always concerned with the realization of some action, while epistemic modality is concerned with the expression of the speaker's view of the degree of a proposition's truth. In examining the features of Japanese sentences, we find that epistemic modality is expressed by modals that constitute modal content, while deontic modality is expressed in the propositional content. There is common ground, however, in that in both epistemic and deontic modality the degree of possibility and necessity is always judged from the speaker's viewpoint, which may or may not correspond to an actual event. Even if a speaker commands an interlocutor to carry out an action and the utterance itself expresses strong necessity, the proposition itself does not reveal whether or not the action will be actualized.

The results of examining the behavior of Japanese modals clarify the idea that modality is expressed in the modals to a certain degree, and that epistemic concepts govern the central notion of modality. In this central notion of modality, as the degree of necessity increases, a proposition comes more and more to resemble a simple factual statement. The reason modals are used by a speaker is that the speaker wishes to append his/her opinion or judgment to a statement in order not to appear to assume that the statement is always necessarily true. If the statement is absolutely and logically true there is no necessity for the speaker to append a modal. When one conveys a known fact, modal content is an unnecessary element. This is why, for the most part, a sentence in the past does not express modality. Historical facts exist as truth until they are overtly questioned. This is shown in the use of the past tense form of some modal *hazu.* When a speaker uses the past tense form of *hazu (hazu dat-ta),* it no longer functions as a modal, that is, it belongs in the domain of a proposition, meaning that "something has been decided, but it did not work out the way it was expected," and it is not related to a speaker's supposition. Thus, epistemic modality involves speaker-oriented notions of possibility and necessity that are not expressed through deontic modality.

Modality then is a matter of the degree of possibility and actuality of a not-yet-realized proposition, that is, of possible/nonactual world situations. The relationship can be conceptualized such that when the degree of possibility and actuality decreases, the degree of modality increases. On the other hand, if there is a high possibility that an event will be actualized, there is a low degree of modality. The notion of necessity becomes secondary to the notion of possibility. This supports Lyons' assertion that "possibility rather than necessity should be taken as primitive in the analysis of epistemic modality" (1977, p. 802).

Defining modality as a speaker's psychological attitude, a definition recognized by most researchers, is certainly all-encompassing. Within that definition, however, it is helpful to approach modality from the perspective of possible/nonactual world concepts, which also correlate with the notions of indeterminability and hypotheticality. On the basis of this approach, modality can be seen as present not only in syntactically modal forms, but also in propositional content, as exemplified by conditional sentences. Thus, by refocusing the definition of modality and approaching its study from a different perspective, the degree of modality expressed by various modals and the presence of modality in both modal and propositional content are better understood.

CHAPTER 4

Modality and Aspectual Categories

This chapter discusses aspectual categories (progressive/resultative [perfect] *te-iru,* and negative *nai*) in relation to propositional content and considers the possibility/impossibility of their expressing modality. It also provides a brief literature review concerning the development of the study of aspect in both English and Japanese. Finally, it proposes refinements to current views of aspect in Japanese by incorporating a new perspective based on the study of English.

Based on the definition of modality outlined in this book, it is immediately apparent that *te-iru,* which involves propositional content, lies outside the scope of modality, attributable to the fact that *te-iru* is used to express a realized event which is a progressive or a resultative situation. Therefore, the event is no longer in the domain of possible/nonactual worlds. The negative suffix *nai,* however, requires further examination, since *nai* can be used in the internal negation structure (in the domain of propositional content) or in the external negation structure (in the domain of modal content). The latter follows the same syntactic behavior as modals, especially *-no-da* and *wake-da.* Although the syntactic behavior of *nai* has been the subject of inquiry, it has not yet been approached from the viewpoint of modality as outlined herein.

Reexamination of aspectual categories is first given to justify the premise that *nai* should also be recognized as an aspectual category by examining the behavior of both *te-iru* and *nai*, a discussion followed by an examination of the possibility/impossibility of *nai* to express modality.

4.1 THE DEVELOPMENT OF THE STUDY OF ASPECT IN JAPANESE

Aspect refers to the temporal property of events and is expressed by various temporal structures such as terminative, durative, semelfactive,[1] iterative, habitual and so forth. In the Japanese language, almost a century has elapsed since the first studies of tense and aspect. Examination of aspect by divorcing it from tense had been considered difficult in part due to the agglutinative feature of Japanese, and the study of tense originally took center stage, with the examination of aspect occurring within that study. The study of aspect became more visible and unchained from the study of tense when Matsushita (1928) established terms for the auxiliary verbs, such as *te-shimau* (completion of action), and *te-iru* and *te-aru* (realized situation) as independent categories.[2] Matsushita claimed that tense is a deictic category that locates situations in time, generally with reference to the present moment, though also with reference to other situations.[3] On the other hand, forms like *te-iru* are concerned with situations that express either

1. The term semelfactive was introduced by Comrie (1976), who defined it as referring to "a situation that takes place once and once only" as opposed to "iterative," which is "a situation that is repeated" (p. 42). The use of semelfactive hereafter is based on Comrie's definition.
2. *Shimau* in the *te-shimau* construction itself means "to put things away." When *shimau* is attached to the *te* form of a verb, however, the compound verb means "completion of an event," which indicates an unrecoverable situation. *Te-shimau* therefore extends the meaning to indicate that the speaker regrets that the event described by the verb cannot be recovered to the original state. *Te-iru* indicates that an event has already been realized. It can express such meaning by referring either to a progressive or to a resultative state, such as "I am eating now" or "I have eaten." Both of these sentences are expressed as "*Tabete-iru*" in Japanese. The function of *te-aru* is similar to that of *te-iru* in that *te-aru* is also used to describe a resultative state—"the state that is a result of the action denoted by the verb to which *aru* is suffixed" (Tsujimura 1991, p. 98). The difference is that *te-aru* is used when the resultative state was achieved through the intention of an agent. Therefore, *te-aru* accompanies only an accusative or an unergative verb.
3. Prior to this work, Matsushita (1901) had also defined tense in terms of absolute tense and relative tense.

a progressive or resultative aspect of an event. Divorcing the aspectual categories from the tense categories thus had begun to be seen in Matsushita's work, and since then, various scholars such as Sakuma (1936) and Miyata (1948) had examined the lexical meaning of those auxiliary verbs with focus on their aspectual characters. To later scholars, aspect theory came to be an independent area of study and was researched morphologically as well, namely by Okuda (1984). The study of aspect has since become a prevalent topic. Most scholars now accept that Japanese has tense, and many are concerned with defining the distinction between tensal meaning and aspectual meaning as manifested in the syntactic behavior and the lexical meaning of verbs.

The study of aspect evolved largely through the work of Kindaichi (1976a), who isolated the concept of aspect in Japanese verbs as opposed to tense, and classified verbs into four categories from an aspectual viewpoint, that is, the lexical meanings of verbs were examined from an aspectual viewpoint. The four categories include: stative verbs that do not accompany *te-iru;* instantaneous verbs that express a resultative meaning when accompanied by *te-iru;* continuative verbs that express either a progressive meaning or a resultative meaning with *te-iru;* and type 4 verbs that always accompany *te-iru* and express a stative meaning.[4] Later on, this classification would be challenged based on the criticism that it was based merely on the idea that the meanings of verbs can be defined by quantifiable units of time. Although this problem was successively reexamined and underwent a number of modifications by various scholars, Kindaichi's original work made such a monumental imprint on this study that for later scholars, namely Mikami (1963), Suzuki (1957), Takahashi (1969), Teramura (1971), and Yoshikawa (1976), his work often served as the point of departure for inquiry, regardless of which viewpoint the linguistic argument would adopt. Even in the modern study of aspect, the argument regarding verb categories, at times, is treated no better than by Kindaichi, as seen in Machida (1989).

Kindaichi's classification of verbs based on their meaning was derived from the idea that an event is a measurable unit. Such a premise was the subject of many reexaminations by a number of scholars, most notably by Okuda (1984), whose most valuable contribution to the study of aspect in Japanese is his proposition regard-

4. Representative verbs for these four categories are: aru "to exist" (stative verb); *tsuku* "to arrive" (instantaneous verb); *nomu* "to read" (continuative verb); and *sugureru* "to excel" (type 4 verb).

ing the contradiction caused by Kindaichi's theory. He evaluated Kindaichi's verb classification in terms of the transitivity and intransitivity dichotomy of Japanese verbs and argued that the use of punctual and durative classification is an unsatisfactory basis for aspectual categorization. For example, transitive and intransitive verb pairs such as *Denki o tsukeru*, "To turn on the light," and *Denki ga tsuku*, "The light will turn on" receive progressive and resultative interpretations, respectively, when *te-iru* is attached.[5] Nevertheless, both of these expressions involve the same expanse of time. Also, based on Kindaichi's classification, punctual verbs express a resultative meaning when attached to *te-iru*, but verbs such as *futoru* "to gain weight" and *tsukareru* "to become tired" can hardly be considered as expressing punctual events. Conversely, the verb *matataku*, "to blink," which can be regarded as a punctual event, does not express a resultative meaning, but rather a progressive/iterative meaning when combined with *te-iru*. The punctual/durative dichotomy is thus not a viable method for classifying Japanese verbs.

Okuda proposed that the process of classifying verbs should not be based on time expended in an event, but rather on whether or not a change in state takes place in the subject. This analysis provides us with two kinds of subjects: a subject that undergoes a change, as is the case with the subject of *futoru*, "to gain weight," and *shinu*, "to die," and a subject that performs an action, as with *tsukeru* "to turn on," and *matataku*, "to blink," for example. This dichotomy is generally represented in Japanese as one of intransitivity versus transitivity: when attached to *te-iru*, intransitive verbs mostly receive a resultative (perfect) interpretation and transitive verbs, a progressive/iterative interpretation. The notion of change can no longer be captured in terms of an expanse of time; instead, the change itself must be considered, even though a certain degree of arbitrariness in determining the point of change is unavoidable depending on the type of event. Verbs such as *futoru*, "to gain weight," *shinu* "to die," and so forth indicate physical and/or psychological changes in the subject

5. In Japanese, intransitivity and transitivity are represented by a different verbal morpheme. There are no unified morphological rules for the formation of transitive and intransitive verbs: some transform causative forms, some attach *eru*, and others transform *eru* to *su*, and so forth. For example, in English, "to turn" can function as either intransitive or transitive. In Japanese, however, this distinction is represented by the different verbal morphemes *mawaru* and *mawasu*, respectively. In this section, a detailed discussion of intransitivity and transitivity is not provided. For a detailed discussion, see Miyagawa (1989), Tsujimura (1991), Jacobsen (1992b), and Kageyama (1996).

that occur over intervals of time of varying length. This point is the source of considerable controversy in the English study of aspect, particularly as undertaken by Dowty (1979).

4.2 THE DEVELOPMENT OF THE STUDY OF ASPECT IN ENGLISH

It is intriguing to note that a similar development had also been occurring in the study of aspect in English. Specifically, another tradition of aspectual study, the roots of which can be traced to Aristotle, was exemplified in the research of Vendler (1957). He proposed a quadripatition of verb categories: states (ongoing in time, e.g., "love" and "know"); activities (ongoing in time, but distinguished from states by a variety of tests, e.g., "walk" and "push a cart"); accomplishments (having some duration and a definite terminus, e.g., "paint a picture" and "run a mile"); and achievements (also having a definite terminus, but occurring instantaneously, with little or no duration, e.g., "discover" and "reach the summit").[6] Verbs were classified into these categories based on whether or not they could accompany the "ing" form and whether or not they could accompany either "for how long" or "how long did it take to."[7] Jacobsen (1982) examined closely the

6. The descriptions provided in parentheses are from Tenny (1994, pp. 4–5).
7. Vendler's chart:

	How long did it take to ~	For how long did it ~	~ing	Occur at a single moment	Unique and definite time
Activities walk, run swim push a cart	-	+	+	-	-
Accomplishments paint a picture draw a circle run a mile	+	-	+	-	+ (imply)
Achievements find, discover reach the summit recognize a man	+	-	-	+	+ (imply)
States desire, want, love believe, hate know	-	+	+	-	-

similarity between Vendler's study of aspect and Kindaichi's verb classification and pointed out that both researchers rely heavily on a definition of the meaning of verbs in terms of how long it takes for an event to occur as a basic criterion for categorizing verb types. Although Kindaichi's classification does not precisely correspond to that of Vendler, the ideas behind such verb classification suggest a methodological, yet some might say nonlinguistic, alteration of linguistic inquiry that may have been inevitable during the course of linguistics history.

Vendler's verb classification was modified shortly thereafter by Kenny (1963), who noted that both accomplishments and achievements can be questioned by "how long did it take for one to ~," but neither can be questioned by "for how long did one do ~." Furthermore, he observed that there cannot be an accomplishment without a closely related end-point achievement. Hence, Kenny proposed three categories: activities, states, and performance, the last of which consists of Vendler's accomplishments and achievements. A number of scholars, such as Taylor (1977), Mourelatos (1978), Dowty (1979), Carlson (1981), Hoeksema (1985), Verkuyl (1993), and Tenny (1994), followed by pointing out insufficient notions and categorical problems recognized in Vendler's (and his other followers) categorization and attempting to establish a theory of aspect on more linguistic grounds. They further built upon Vendler's classification but, despite their pointed criticisms, do not seem to have succeeded entirely in repainting the picture Vendler had drawn. Their works are significant, however, in that the focus of the argument is not in determining the duration of an event (i.e., viewing verbs in terms of a measurable unit) but in viewing the aspectual property from an interaction of syntax and semantics.

Among the theories presented by various linguists, Dowty's (1979) should be recognized as one of the most detailed and rigorous grammars of the English verbal system. He pointed out vulnerabilities in Vendler's work and built upon his verb classification by revising it and introducing a distinction between definite and indefinite changes of states. There was some ambiguity in the distinction Vendler made between accomplishment and achievement verbs. Achievement verbs supposedly do not assume the "ing" form, but there are some examples counter to this observation, such as "he is dying," or "he is falling asleep." Dowty approached this problem in terms of "interval semantics," which further examines Vendler's four categories through the use of operators such as "DO," "BECOME," and "CAUSE." For example, as shown in Verkuyl (1993), if Vn is an *n*-place predicate and $a_1, \ldots, a_n$ are its argument(s), the four classes are approximately represented as follows:

(1)

States	$V_n(a_1, \ldots, a_n)$	John knows the answer.
Activities	DO $(a_1, V_n [a_1, \ldots, a_n])$	John is walking.
Accomplishments	DO $(a_1, V_n [a_1, \ldots, a_n])$ CAUSE BECOME $(Vn[a_1, \ldots, a_n])$	John broke the window.
Achievements	BECOME $(V_n[a_1, \ldots, a_n])$	John discovered the solution.

The "do" operator indicates that the interval described by the proposition is true in an interval if and only if it is true in all subintervals. This operator alone, however, does not account for accomplishments and achievements that hold for definite changes. Accomplishments require that the operators "DO," "CAUSE," and "BECOME" go together, while achievements require only one operator—"BECOME." Accomplishments are therefore labeled [+Complex], while achievements are labeled [–Complex] in order to distinguish the characteristics of these two types of categories.

Dowty also used "for-adverbials" and "in-adverbials" to analyze sentences such as "John slept" and "John awakened."[8] He explained that states and activities are "for-adverbials" since both predicates are true for most or all subintervals of an interval of the predicate's truth, while "in-adverbials" are compatible with accomplishments and

8. Dowty explains that for-adverbials are appropriate for states and activities since both classes of predicates are true of most or all subintervals of an interval of the predicate's truth, but not for accomplishments and achievements since they are nonsubinterval predicates. "In-adverbials" are appropriate for accomplishments and achievements since they can satisfy the "uniqueness" requirements in the semantics of in but not for activities and states. Note the following examples:

(a) John slept for an hour.
(b) ?John slept in an hour.
(c) *John awakened for an hour.
(d) John awakened in an hour.

achievements since both can satisfy the uniqueness requirements in the semantics of "in." He examined two of Vendler's four categories,[9] accomplishment, and achievement together with the operators mentioned above, and reproduced them into two categories that Dowty referred to as single change of state [–Complex] and complex change of state [+Complex]. This revised verb classification then deals with states and activities as nonchange predicates, and single changes of state and complex changes of state as change-of-state predicates. Nonchange predicates involve indefinite changes of state, and change-of-state predicates involve a definite change of state, the two types falling into the homogeneous and heterogeneous categories, respectively. Corresponding terminologies, such as telic-atelic by Dahl (1981), count/duration by Hoeksema (1984), culminated/nonculminated by Moens and Steedman (1988), bounded/nonbounded by Jackendoff (1990), and delimited/nondelimited by Tenny (1994), were later proposed in succession, with some different issues of compositionality with each other.

Dowty's theory on homogeneity is eventually commented upon by Verkuyl (1993), who approached aspectuality from the interaction between atemporal and temporal structures of the language. Verkuyl points out that Dowty's theory "focuses too much on the construc-

9. Vendler's "accomplishments" were typically from categories 6, 7, and 8, and "achievements" mainly from category 5.

	Non-agentive	Agentive
States	1a. be asleep be in the garden love, know	2a. be polite be a hero
Activities	1b. interval statives: sit, stand, lie	2b. interval statives: sit, stand, lie (with human subject)
Single change of state	3. make noise, roll, rain 5. notice, realize, ignite	4. walk, laugh, dance 6. kill, point out
Complex change of state	7. flow from X to Y dissolve	8. build a house, walk a mile walk from X to Y

tion of temporal domain structure rather than on the question of how language can be seen as a means to construe structure by the choice of the constituents that make up a sentence (p. 67)."[10] In other words, Dowty's theory does not account for the difference between a sentence with a singular indefinite direct object NP such as "I ate an apple" and a sentence with a quantified NPs such as "I ate apples".

Although linguists like Verkuyl make the objection that Dowty's analysis is a "holistic approach and focuses on a restricted class of lexical predicates" (p. 206), and that it offers no difference from traditional approaches that focus on just verbal properties,[11] homogeneity did play a central role in the study of aspect in the 1970s. Verkuyl considered his theory a "top-down" approach that observes an interval, with a beginning-point and ending-point, and asks questions about its internal structure. Such a theory is quite different from a "bottom-top" approach that builds up a structure and results in one having a beginning-point and an ending-point separated by a certain duration of time.

Although there are arguments in favor of and opposing the theory of aspect proposed by Dowty, his theory provides new insight into the study of aspect in Japanese. Thus, the following examination is based on the top-down approach consistent with the work of Dowty.

4.3 FUNCTIONS OF THE ASPECTUAL FORM *TE-IRU*

When reviewing Dowty's theory of aspect, one immediately notices the striking differences between English and Japanese in the way homogeneity and heterogeneity are captured by the property of the aspectual form *te-iru*. In Japanese, the central focus of the study of aspect has been mainly placed on the study of the auxiliary verb *te-iru*, which is a barometer of categorizing the lexical aspect of verbs. Jacobsen (1984) introduced Dowty's reductionist approach into the study of aspect in Japanese. He interpreted Dowty's lexical homogeneity as stating that "a predicate is homogeneous if and only if for any interval of time over which a statement consisting of that predicate and its associated noun argument is true, the statement is also

10. For example, the "become" operator may not be applicable when a predicate accompanies a different complement, i.e., Dowty did not pay attention to verbs that may or may not be bounded by the particular choice of their complements.

11. For a detailed discussion, refer to Verkuyl (1993, pp. 201–14).

true for any subinterval of that interval, no matter how small" (p. 151). Jacobsen suggested that the fundamental concept of aspect can be approached from a unified analysis of the notion of homogeneity as exemplified by the functions of the Japanese aspectual form *te-iru,* which is capable of expressing either progressive or resultative (perfect) aspect. The theory of homogeneity as a top-down approach not only applies well to the Japanese aspectual form *te-iru,* but also broadens the scope of aspectual categories viewed by the notion of homogeneity.

4.3.1 Homogeneity and *Te-iru*: Broadening Aspectual Categories in Japanese

In the study of aspect in Japanese, researchers have emphasized the developmental stage of a situation expressed by verbal predicates. This gives the impression that the study of aspect concerns only verb categories. As was shown, however, *te-iru* is used to express both perfect meanings (states) and progressive/iterative meanings (activities), both of which are in the homogeneous category. States and activities are grouped together into one category sharing the common feature that a statement which is true of any interval is also true of various subintervals of that interval. Therefore, if *te-iru* is labeled as an aspectual form, then stative predicates, which are in the homogeneous category, should also be recognized as having aspectual meaning.

Japanese stative predicates are represented by stative verbs such as *aru,* "to exist," *wakaru,* "to understand," *dekiru,* "to be able to do," adjectives, the copula *da,* and auxiliary verbs that express a resultative aspect such as *te-iru,* and *te-aru,* for example. Note the following examples:

(2) a. Mario wa ima terebi o mite-iru.
he TOP now TV ACC watch-PROG
"Mario is watching TV now."

b. Mario wa kyoo wa moo takusan terebi o mite-iru
Mario TOP today CONT already a lot TV ACC watch-PERF
"Mario has already watched TV a lot today."

(3) a. Kodomo ga suyasuya to nete-iru.
child NOM peacefully sleep-PROG
"A child is sleeping peacefully."

b. Uchi no ko wa kyoo wa takusan nete-iru.
my GEN child TOP today CONT a lot sleep-PERF
"My child has slept a lot today."

(4) a. Sensei wa ushiro no seki ni suwatte-iru.
teacher TOP back row GEN seat LOC sit-PROG
"The teacher is sitting in the back row seat."

b. Kono seki ni wa mae ni mo suwatte-iru.
this seat LOC TOP before too sit-PERF
"I have sat in this seat before."

(5) Okashii naa . . . Heya no denki ga tsuite-iru.
strange PART room GEN light NOM turn on-PERF
"Strange . . . The lights in the room are on."

(6) Kanako wa katei no shufu-da.
Kanako TOP family GEN house wife-COP
"Kanako is a housewife."

(7) Saikin no terebi geemu wa jitsuni omoshiroi.
recent GEN TV game TOP really interesting.
"Recent video games are really interesting."

The homogeneous predicates in sentences (2)–(4) accompany the aspectual form *te-iru,* yet the meaning differs due to the employment of adverbial phrases, suggesting that the event is in a progressive or resultative state. Without such employment, the meaning is ambiguous. *Tsuku,* "to turn on," in sentence (5) is, however, a heterogeneous predicate and therefore offers only one interpretation—resultative—

as was exemplified in the previous section on heterogeneity. Accompanying *te-iru*, therefore, means a conversion of events into the homogeneous category, since *te-iru* itself represents homogeneity. These events share a common feature in that any subinterval of the interval in question can be recognized as having the same property. For example, any subinterval of the state of "watching TV" is watching TV, in the same way that any subinterval of a resultative state, "has watched TV," for example, is having watched TV. This observation can also be applied to sentences (6) and (7) where stative predicates (nominal-*da*, and adjective) are used. For sentences (6) and (7), any subinterval of the state of "being a housewife" and "interesting" is being a housewife and interesting, respectively, no matter how small the subinterval.

It should also be recognized, however, that these activity predicates (activities) and stative predicates (states) do not behave in exactly the same way. As Dowty pointed out, activities are not purely homogeneous in that activities may not be determined at a single instant, like a frozen movie frame, whereas states can be captured at successively smaller intervals even if reduced to a single point. Therefore, states and activities can be represented, for example, by a solid line and a dotted line, respectively. For an activity to be represented by a finely dotted line, a dashed line, or perhaps a charcoalesque line depends on how frequently and how much energy input is required to maintain the activity. This point will be clarified in the following sections as the functions of *te-iru* are discussed in more detail.

4.3.2 Heterogeneity and *Te-iru*

Contrary to homogeneous predicates, heterogeneous (change of state) predicates indicate a change in a subject at a single moment or interval, while suggesting that such intervals are located on a time continuum that reflects different stages of the subject up to and through the change addressed by the speaker. Although the time frame differs from predicate to predicate, each distinct instant exhibits a gradual development that distinguishes it from previous instants in a given time frame. A change in a subject can be thought of as a phenomenon of gradual "rearrangement of things or properties in space" (Tenny 1987, p. 77) from instant to instant over time.

This gradual change from instant to instant over time can be expressed by the progressive form in English, such as "He is falling asleep," "She is dying," "The airplane is arriving at the airport," and "He is reaching the summit." The duration between the time when he starts falling asleep and the moment that he falls asleep, for example, is considered as process in which the event reaches a point of change. The striking difference between English and Japanese here is the fact that this process of the event reaching a point of change cannot be expressed by the progressive form in Japanese. Heterogeneous predicates fitting into Vendler's "achievement" and "accomplishment" categories, including those falling under Kindaichi's "instantaneous" category, express a resultative (perfect) meaning when *te-iru* is attached. As was pointed out by Okuda, these instantaneous predicates do not necessarily depend on the immediacy of the event to occur; rather, they depend upon whether or not a change is brought to a subject. Jacobsen (1992) also stated that these predicates lack a homogeneous quality, and "the transition involved between such differing states of affairs is unique in that it cannot be arbitrarily predicated of any point within a given interval of time" (p. 181).

Most intransitive verbs (unaccusatives) are the representative case of the above description as they assign an entity undergoing a change to the focal subject role.[12] The following are examples of the intransitive verbs *shinu*, "to die," and *yaseru*, "to lose weight," in Japanese:

(8) a. Kono ueki wa shinu.
this plant TOP die
"This plant will die."[13]

b. Han'nin wa sono jiko de sokushi-shi-ta.
criminal TOP that accident INS instant death-do-PST

12. The term "intransitive" in this context refers to "unaccusatives" in the sense familiarized by Perlmutter (1978) and Levin (1995). In this book, accusativity, unaccusativity, and unergativity are, however, outside the scope of examination.

13. The best way to express the current progress in the situation "The plant is dying" may be the use of "*-tsutsu aru.* (is in the process of ~)." This phrase, however, is seldom used in colloquial conversation.

"The criminal died instantly in the accident."

(9) a. Kekkkon shi-ta totan ni 5 kiro yase-ta
marry-PST just at the moment 5 kiro lose weight-PST
"I lost five kilograms as soon as I got married."

b. Kono kusuri o nomu-to, 2 kagetsu de 10 kiro yaseru.
this pill ACC drink-COND 2 months in 10 kiro lose weight
"If you take these pills you can lose ten kilograms within two months."

Sentence (8)a is describing a situation, "The plant will die," meaning "the plant is dying" in English, where a plant is perceived as changing from the state of being alive to the state of not being alive. The plant's condition will worsen in its internal and external appearance over each instant of time as it moves toward biological death. Each instant does not display exactly the same aspect of the plant. That is, each instant cannot be described as a repeated event or phenomenon, but the event as a single whole presents a semelfactive event[14] in Japanese. In sentence (8)b death occurs much faster than in sentence (8)a in terms of time expanse. Nevertheless, when viewing this event on an imagined time line, if we consider the time line to be a kind of filmed record, we can isolate each frame of that time line for theoretical purposes. Whether the "instant" is captured by 10 frames, 100 frames, or 1000 frames, it is possible to notice a gradual change in an event from frame to frame until the entity becomes not alive (expressed by BECOME ∅). This imaginative time line incorporated with the event can be illustrated in the following diagram:

14. Dowty's "single change of state" predicates and Kindaichi's "instantaneous" predicates are typically semelfactive since those events cannot occur iteratively within the interval with the end point being the change of state. *Shinu* "to die" is the most conceivable semelfactive event since it is brought only once to a subject. *Kizuku* "realize" and *futoru* "to gain weight" are also semelfactive events. Although these events can be repeated, the occurrence has to be once per occasion.

(10)

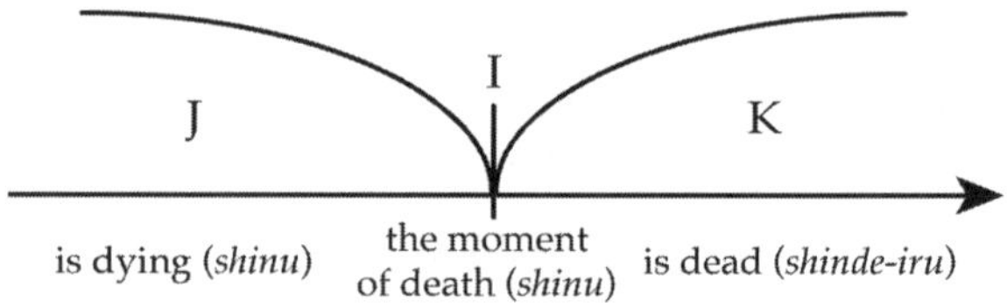

This illustration shows that there is an interval "J" containing the bound of "I" such that →∅ is true at "J," and there is an interval "K" also containing the bound of "I" such that f is true at "K." Figure (10) can also apply for sentences (9)a and b. To become skinny means to lose a certain amount of body fat, water, and/or muscle, for example. These constituents disappear as time passes, and this brings about a change from the original appearance of the subject at the time of speech. No matter how fast one loses weight the same developmental process can be seen, that is, a reduction in mass of the elements that constitute the human body. The problem, however, is the difficulty in establishing a single point of change depending on the type of event, since an achievement cannot occur over every subinterval of a given interval, but can occur over at most one single interval, with this single interval conceptualized as the minified or magnified moment of change.

This single interval, in fact, becomes a problem with Dowty's BECOME operator for those who approach his theory in terms of "truth condition." It is obviously the case that the change represented by the verb "to die" and the verb "to become skinny" cannot be perceived in exactly the same fashion; it is difficult to recognize the moment of change in the event "to become skinny." Galton (1984) also recognized this point and stated that a change is not necessarily instantaneous, but some changes take time. Dowty acknowledged this problem and proposed a two-place operator for change, which deals with the interval "I" located between the durations "J" and "K" with a condition that there is no nonempty subset of "I" such that a and b hold for "I^A" and "I," which means that the interval "I" after all represents a concept that the process of becoming must be abridged to two moments. In this regard, Galton (1984) made a similar observation where he takes the example of a banana's ripeness. When a banana is said to ripen, it is inevitable that it pass from a state in which it is clearly unripe to a state in which it is clearly ripe. Galton claimed that there exists an interval "grey area" within which there is a series of states that are not clearly states of ripeness nor states of unripeness. The following chart represents the notion of "I" and Galton's grey area.

(11)a

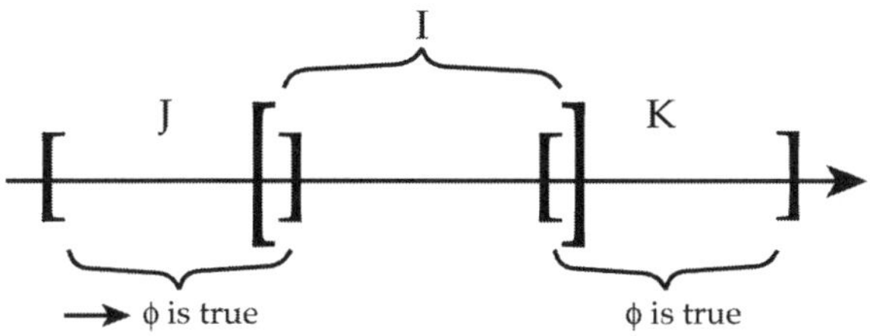

Such identification of an interval "I," however, seems to make the situation appear even more difficult, especially for events that take time to change, such as "to lose weight" and "to cool down." It might be possible to say that death, for example, can be measured in terms of a quantifiable time unit, such as brain death taken as a lower bound, followed by heart death, or vise versa, depending on how death is medically, legally, and/or practically defined. For example, if brain death is defined as the legal point of death, then the moment between brain death and subsequent heart death must be perceived as "death," holding the same temporal quality that is seen from the point that death is realized. What then, with regard to examples such as "lose weight" or "gain weight?" In real life, no matter how much we wish it were so, time cannot be frozen to check its internal quality.

What is the quality of the interval "I"? Is it similar to the interval "→∅" or is it identical to the initial point of "∅ " as the resultant from losing weight? In either case, how does this gap relate to temporal advancement? The interval "I" can only be conceptualized. Dowty tried to apply a pragmatic principle to defend this interval. As pointed out by Lascarides (1988), however, this proposal is at odds with the notion that Dowty's homogeneity should be consistent in quality in any subinterval of the interval, and should not include a different quality for the interval "I." This critique suggests that the time approaching the point of death and the death of an entity progressing should hold the same time expanse; furthermore, there should not be any temporal gap between the point that death is realized and the state after the occurrence of death. Dowty's pragmatic principle is an unavoidable imposition, since, after all, every natural language phenomenon cannot be treated mathematically.

Within the realm of pure mathematics, of course, the notions of "grey areas" or "intervals" are largely unnecessary. For example, well-defined mathematical functions (e.g., the tangent or step function) may change infinitely abruptly (i.e., discontinuously) from one state to another. In the case of the tangent, for example, the function changes from infinitely positive to infinitely negative at odd multiples of pi. In the case of the step, the function changes discontinuously from a value of zero to a value of one at a single point. In neither case does a "grey area" exist between the two domains of the function, and the value of the function—even at the point of transition, provided that it is defined—may be evaluated by calculating the limit. Discussions of the problems surrounding "grey areas" and "intervals" smacks somewhat of the quandaries surrounding Zeno's Paradox, which posed a philosophical enigma regarding even the simplest notions of physical reality. Zeno argued that even Achilles could never run from point "J" to point "K," since to do so, he would first have to reach a point half-way between them. But to reach that point would require him first to reach a point one-fourth the distance from "J" to "K." But to reach that would require him first to reach one-eighth the distance, which requires him first to reach one-sixteenth the distance, and so on. Therefore, Achilles must cross an infinite number of intervals (each of which has finite length) in a finite amount of time; hence, he could never reach "K." In fact, argued Zeno, he could never even leave point "J," and Achilles—and all the world—is by logic constrained not to move.

Within the realm of physical reality, on the other hand, one's ability to objectively assess the state of a situation depends upon one's ability to measure it, that is, to observe it either directly or indirectly. To identify a situation as "J" at a time, t_1, and as "K" at a later time, t_2, is well and good provided that one has measured the state of the situation as such at times t_1 and t_2. However, to discuss the state of the situation at an intermediate time—that is, within the "interval" between t_1 and t_2—is to speculate beyond the scope of objectivity if the state was not measured, but merely inferred or presumed. For example, even if one measures "J" at t_1 and "J" again at t_2, how can one be sure that the situation was not "K" at some moment in the interval? And if one measures "J" at t_1 and "K" at t_2, then to characterize the point of transition would require higher and higher degrees of temporal resolution. Ultimately, however, meaningful observations within the realm of physical reality are limited to a finite degree of temporal sampling. Hence, it is not the "interval" of a transition between states "J" and "K" but strictly the concept of the transition and the fact that a transition occurred that distinguishes one state from the other. The states are separated by an unmeasured domain—herein referred to as the "Boundary"—within which it is meaningless and speculative to identify the exact nature of the situation. This boundary coincides with the occurrence of change or with that which brings a change to a subject.

The "I" problem in heterogeneous predicates is at least structurally avoidable in Japanese, since the interval "J" and "I" can be represented by the same verbal expression when the event is perceived as a neutral description of the event.[15] Conceptually, we can envision death, for example, as a course of events in progress, no matter how short or long that course may be to reach the final point. That is, "something is dying" represents a change that occurs via a step-by-step progression, irrespective of the length of the interval required. This, however, is not reflected in the way the Japanese language is used.[16] For example, in English, the time table can be described in three different structural patterns, "is dying," "become dead/die," and "is dead." The interval "I" is equivalent to the second phrase. This step-by-step progression, however, cannot be expressed by the progressive aspectual form in Japanese, unlike that of English, "is dying." No stage of the development procedure can be perceived as the repeated event; rather, the accumulation of such stages taken as a whole constitutes the change. Therefore, the first two intervals are represented by *shinu*, "to die,"[17] and the third, *shinde-iru*, "is dead," describes the event in general. The transition

15. If, however, the event in question is a past particular event, the interval "J" cannot even be described by using the heterogeneous predicate past tense form. The expressions such as *shi ni mukatte-i-ta* "~ was heading for death" are used instead.

16. As noted at the end of section 4.2, the discussion does not exceed Dowty's study. That is, lexical homogeneity is the object of the discussion, and aspectuality in terms of the Verkuyl approach (e.g., examining the difference between "Judith ate a sandwich," "Judith ate sandwiches," "Judith ate bread," and "Judith ate no sandwich.") is not integrated.

17. Again, when the process is emphasized, some other structural patterns may be used, as exemplified in footnote 13.

from "alive" to "not alive" is defined by the "boundary," which indicates a separation between states. This is simply a conceptualized, pragmatic separation and can be perceived as time-based or as sense-based depending on the particular event, and it is not established in terms of formal semantics. Some events, such as *shinu*, "to die," and *tsuku*, "to arrive," can be recognized at a point in time, while others such as *futoru*, "to gain weight," and *yaseru*, "to lose weight," cannot be recognized at a discrete moment in time. These events are recognized depending on a speaker's perception. The following diagram shows these equivalents in Japanese in the form of the previous chart (11)a:

(11)b

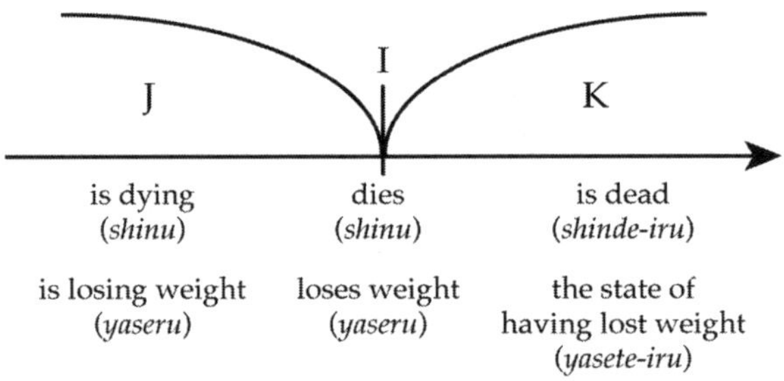

As seen in the chart, when *te-iru* is attached to these heterogeneous predicates, they express only a perfect meaning—state "K," the subject's new position relative to other pertinent things or properties in space. Note the following examples:

(12) Kono ueki wa shinde-iru.
this plant TOP die-PERF
"This plant is dead."

(13) Miwa wa kekkonshi-te, yase-ta. Sore irai yasete-iru.
Miwa TOP marry-GER lose weight-PST since then lose weight-PERF
"Miwa got married and lost weight. Since then, she has been skinny (she is in the state of having lost weight)."

Both sentences (12) and (13) express a perfect meaning indicating that the plant is in the state of having died, and Miwa is in the state of having lost weight, respectively. When a change occurring in a subject can be defined as a semelfactive event, the *te-iru* sentences containing the subject do not suggest a developing course of events. This means that *te-iru* cannot function to express a semelfactive event in progress. For example, *iku* "to go" is one of the intransitive verbs whose functions vacillate depending on a complement such as "(go) to somewhere" and "(go) on this street" as exemplified in the following:

(14) a.

Miwa	wa	hitoride	Sendai	e	itte-iru.
Miwa	TOP	alone	Sendai	LOC	go-PERF

"Miwa has gone to Sendai alone."

b.

Miwa	wa	kono michi	o	massuguni	itte-iru.
Miwa	TOP	this street	ACC	straight	go-PROG

"Miwa is going straight on this street."

In sentence (14)a, the activity "to go" is performed within a limited range whose boundary is Sendai, and which receives only the perfect interpretation. The activity "to go" in this sentence is a step-by-step movement culminating in the achievement of the boundary of Sendai. Each step is viewed in terms of the distance between Miwa and Sendai, and "to go" cannot be regarded as an iterated activity.[18] This is a semelfactive event where change is brought about in the subject. On the other hand, the activity in sentence (14)b is not seen as occurring

18. In order for a change predicate to receive a progressive interpretation, the subject has to be plural or has to be in a position to repeat the same event. Please note the following examples:

(a)

Aru machi	de	kodomo-tachi	ga	tsugitsuigini	shinde-iru.
a town	LOC	children	NOM	one after another	die-ITER

"Children are dying one after another in Bosnia."

(b)

Miwa	wa	nandomo	Sendai	e	itte-iru.
Miwa	TOP	many times	Sendai	LOC	go-ITER

"Miwa has been to Sendai many times."

These sentences receive a progressive/iterative interpretation due to the plural subjects, and due to the adverbial phrases *tsugitsugini* "one after another" and *nandomo* "many times." Many people's deaths and Miwa's going to Sendai occur in a successive way.

in a limited range, but rather as indicating a type of activity. The subject is indeed repeating an action, in this case moving her feet to go forward on the street. This *te-iru* receives a progressive interpretation. These types of commonplace activities, such as *sanposuru*, "to take a walk," *oyogu*, "to swim," *hashiru*, "to run," can receive different interpretations depending on whether or not an ultimate goal is indicated. Predicates that indicate only a semelfactive event, such as *shinu* and *yaseru*, cannot be regarded as iterative events occurring between one end of an interval and the other end of an interval for a single subject. Compare the following examples that use *te-iru:*

(15) Hikooki wa moo kuukoo ni tsuite-iru.
airplane TOP already airport LOC arrive-PERF
"The airplane has already arrived at the airport."

(16) a. Taroo wa ima kawa de oyoide-iru.
Taro TOP now river LOC swim-PROG
"Taro is swimming in the river."

b. Taroo wa kawa o mukoogishi made oyoide-iru.
Taro TOP river ACC the other side to swim-PERF
"Taro has swum to the other side of the river."

When used with *te-iru* the intransitive verb *tsuku* "to arrive" can express only a perfect meaning, whereas the intransitive verb *oyogu* "to swim" can express a perfect or progressive/iterative meaning depending on context. When *oyogu* has a perfect interpretation, it indicates that a change has been brought about by reaching the boundary of an event. These events can possess a boundary that triggers a change in the subject in a manner similar to a semelfactive event, except that in these a boundary is reached due to the subject's iterative activity upon an object. The object acted upon by the subject will be exhausted as the boundary approaches. For example, "to drink a glass of water" expresses the consumption of the water in the glass. The water will repeatedly and gradually be consumed by the subject until it reaches the boundary of complete disappearance. For other activities where the consumption of the object is not a concrete concept, such as "to take a walk in the park," the object, in this case "the park" (or more precisely the ground in the park), is not consumed, but instead is affected by the subject. Tenny (1987, 1994) has proposed that the notion of change can be approached from "affectedness," which may be defined as the property of a verb such that it describes a phenomenon of "delimitedness." On the basis of her premise, affectedness verbs describe events that are "measured out" and delimited by their direct arguments. If an argument undergoes a change gradually, it measures out the progress of the event toward that change (1987, pp. 77–86). Tenny's proposal accounts remarkably well for the activity "to take a walk in the park" in that the event (to take a walk) measures out the object (park). The key difference between expressions that receive both progressive/iterative interpretations and perfect interpretations, and those that receive perfect interpretations only, is thus in the way an event arrives at a boundary of change. That is, the difference lies in whether it arrives as a semelfactive event or an iterative event.

The dichotomy of homogeneous and heterogeneous predicates is not perfectly analogous to the dichotomy of transitivity and intransitivity, since intransitive verbs include activity verbs (such as *aruku*, "to walk," and *oyogu*, "to swim") that are considered iterative events and in the category of the homogeneous.[19] This dichotomy does not precisely account for aspectual categories giving rise to the different meanings of *te-iru*, which become much more apprehensible when homogeneity is examined in terms of "energy input."

4.3.3 *Te-iru* as an Energy Operator

Development of the theory of aspect in terms of verb categories thus originates in the philosophical cogitations of Vendler. In linguistics, there was another evolution occurring in the traditional Slavistic view of aspect, which examines aspect as a morphological-lexical issue. This work was refined in English by Comrie (1976), who defined aspect as "situation-internal" time, and tense as "situation-external" time. He

19. Although it is not included in the discussion, the verb categories proposed by Perlmutter (1978) correspond to the homogeneous and heterogeneous categories well in Japanese. On the basis of his categories, most accusatives and unergatives are classified as homogeneous predicates, whereas most unaccusatives are heterogeneous predicates.

further provided a definition of terms related to tense and aspect such as "perfective" and "imperfective." Based on Comrie's study, "perfective" indicates a completed action with a beginning, middle, and end but does not necessarily put emphasis on the termination point; rather, all parts of the situation are presented as a single whole. "Imperfective" is defined as an explicit reference to the internal temporal structure of a situation, in which the situation is viewed from within. Given this analysis, perfectivity is in fact compatible with imperfectivity. For example, in "John was cooking when I returned home," the phrase "was cooking" views the situation with regard to its internal structure and "returned" views the situation as a single whole. Comrie also provided a definition of the term "perfect," denoting a past situation that has present relevance as well as the present result of a past event, as in "He has broken his arm," for example. Comrie placed the categories of perfective, imperfective, and perfect in the "aspect" category. In the traditional Western view of aspect, which regards the notion of time as a fundamentally quantifiable category, the dichotomy of perfective versus imperfective is closely related to that of punctual versus durative.

Comrie (1976) also made a similar point in terms of this dichotomy between dynamic and static. In order for an activity to be maintained, new energy has to be continually introduced to the activity. Also, activities are made up of phases that are different from one instant to another. For example, if a given person is eating, that person is involved in a succession of subactivities that maintain the activity (e.g., cutting food, chewing food, and so forth). Conversely, pure states do not require such an effort; they continue eternally unless some kind of outside energy is added. This is also reflected in the fact that activities accompany *te-iru* whereas pure states do not.

Energy in this discussion, however, may be derived not only from animate, volitional subjects, but also from inanimate subjects and natural forces. The latter case is especially seen in the use of *te-iru* in association with personified subjects. For example, "typhoon" itself is not a volitional subject, but we often envision its power in a personified fashion, allowing us to envision the typhoon as a volitional force. *Taifuu ga atarashii tatemono o kowashite-iru,* for example, is interpreted as either "The typhoon is destroying the new buildings" or "The typhoon has destroyed the new buildings." Also, there are cases where *te-iru* receives only a progressive reading. These progressive events are caused by natural forces, but describe simple natural phenomena rather than personifications, such as *Hoshi ga hikatte-iru*, "A star is blinking," *Kaze ga fuite-iru*, "Wind is blowing," *Hata ga kaze ni nabiite-iru*, "A flag is fluttering in the wind," and *Inazuma/kokki ga hirameite-iru*, "The lightning/a national flag is flashing." For such natural phenomena in progress, the subject does not have willful power to change the situation to a resultative one.

From this viewpoint, the basic meaning of *te-iru* can be seen to involve an input of energy to an event by a sentient being that can control the event volitionally and/or intentionally, that is, the basic concept of "agentivity."[20] Resultative (perfect) *te-iru* itself comes about due to an influx of energy that converted some previous state into a resultative state. Consequently, *te-iru* indicates whether energy is being added to a situation or not. This function of *te-iru* in terms of volitional energy input explains why intransitives such as *shinu* "to die," *ki ga tsuku* "to notice," and *tsuku* "to arrive" cannot receive a progressive interpretation, while other intransitives such as *tatsu* "to stand up" and *aruku* "to walk" can receive either progressive or perfect interpretations. That is, neither *shinu* nor *ki ga tsuku* is volitionally controlled, whereas both *tatsu* and *aruku* can be. This view of *te-iru* is also related to whether an event is perceived as semelfactive or iterative.

As previously noted, energy is conceptually input only once for a semelfactive event, whereas energy is input into each segment of an iterative event. An event such as *shinu*, "to die," and *tsuku*, "to arrive," are typical semelfactive events that cannot be perceived as iterative events, and therefore receive only a perfect interpretation. On the other hand, *taberu* "to eat" is an iterative event and vacillates between activity and state depending on the situation when *te-iru* is attached. The iterative event refers simultaneously to internal and external situations, emphasizing not only internal repetition, but also inception and termination. Pure states, however, possess none of these features. For example, the predicates *katei no shufu-da* "is a housewife" and *omoshiroi* "interesting" in sentences (6) and (7) express pure states describing the attributes of the person Kanako and a recent video game, respectively. These predicates do not imply inception or termination of an interval, but rather express that the situation is on a continuum with no clear inception or termination points. These predicates express

20. A detailed discussion of "agentivity" is found in Gruber (1967) and Jacobsen (1997).

everlasting, static situations, not dynamic situations, and they are fundamentally different from activities even though they are considered to be in the same homogeneous category. Sentence (2)a, *Mario wa ima terebi o mite-iru,* can express a perfect meaning, "Mario has watched TV," as well as a progressive meaning, "Mario is watching TV." The former implies that energy had been input for a certain period of time and the latter that energy is still being input by the agent.

These examples imply that the energy can be discontinued via agent volition, thus bringing about a resultative state. These notions are provisionally termed "energy input theory," which indicates that a volitional activity can bring an event into two phases—progressive and resultative states—while a nonvolitional activity brings an event into a resultative state.

The energy input theory operates upon Japanese aspectual categories well, but as can be easily seen, it does not apply to English, where the progressive form does not correspond to iterated energy input. This is due to the fact that "progressive" is essentially perceived as a step-by-step progression regardless of the event being semelfactive or iterative. "Something is dying," for example, conveys the state of achieving the point of death, and the process progresses with the passage of time as a temporal concert.

4.3.4 Summary of the Aspectual Form *Te-iru*

The Japanese aspectual form *te-iru* thus holds the property of homogeneity and functions as an energy operator. These two notions—homogeneity and energy input—produce four combinations under which predicates can be categorized:

(17)

1. Homogeneous + energy input:
   Volitional activities
   (e.g., *taberu* "to eat," *miru* "to watch," *oyogu* "to swim" etc.)

2. Homogeneous + zero energy input (no *te-iru* attachment is necessary):
   Nonvolitional states
   (e.g., *iru/aru*, "exist," *utsukushii*, "be beautiful," *shufu-da*, "be a housewife," etc.)

3. Heterogeneous + energy input:
   Volitional intransitives—*okiru*, "to get up,"[21] *neru*, "to lie down," *noru*, "to ride in," *iku*, "to go," etc.

4. Heterogeneous + nonenergy input:
   Nonvolitional intransitives
   (e.g., *tsuku*, "to arrive," *shinu*, "to die," *kieru*, "to turn off (lights)," *futoru*, "to gain weight," etc.)

The following chart represents a conceptual model analogous to a chemical reaction in which *te-iru* acts upon the four categories of predicate:

(18)

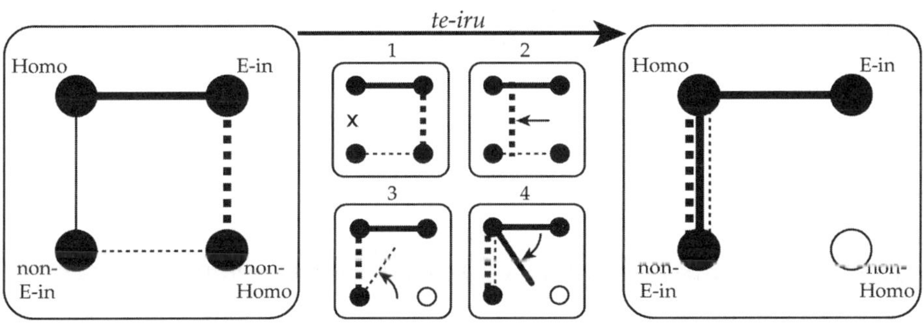

The object (molecule) on the left of the diagram represents the four categories and associated predicates (represented as chemical bonds). *When te-iru* is attached, the effect upon all four categories of predicates (i.e., upon the molecule) can be conceptualized as a series of virtual states, as shown in steps 1–4 (resulting in a new species, shown on the right). As shown in step 1, nonvolitional states that allow neither resultative nor progressive readings are eliminated (i.e., that bond is broken) since they do not allow the attachment *of te-iru*. These events are pure stative predicates that do not require any energy input in order to maintain the state and do not have clear points of inception

21. It should be obvious by this point that a verb like *okiru* can be interpreted as "to wake up," which is a nonvolitional event, and "to get up (from the bed)," which is a volitional activity. The former belongs to the heterogeneous and energy input category, while the latter belongs to the homogeneous and energy input category.

and termination.[22] Thus, the notion of "change" is inapplicable to these pure stative predicates. In this sense, a pure homogeneous predicate that does not subsume a point of change is rather secondary to the aspectual category.

As shown in step 2, volitional intransitives such as *iku* "to go," and *okiru* "to wake up" become nonvolitional states (i.e., the bond between the heterogeneous and energy-input categories is broken, and the bond reforms between the homogeneous and nonenergy-input categories.)

As shown in step 3, nonvolitional intransitives such as *shinu*, "to die," and *yaseru*, "to lose weight" become nonvolitional states that lie between homogeneous and nonenergy input (i.e., the bond with the heterogeneous category is broken, shifting instead to the homogeneous category).

Finally, as shown in step 4, volitional activities such as *taberu*, "to eat," and *oyogu*, "to swim," span over two categories expressing progressive aspect and resultative aspect (i.e., the bond between the homogeneous and energy-input categories splits, and reforms between both energy-input and nonenergy input categories).

As evident in the diagram, heterogeneous events become homogeneous and express a resultative meaning only, since the moment of the event's occurrence cannot be perceived as events in progress. (Conceptually, the product of the reaction is a species devoid of the heterogeneous category.) *Te-iru* as an energy operator thus entices events (other than pure stative predicates) into the homogeneous category, excludes lexical heterogeneity from progressive aspect, and places pure stative predicates into a category secondary to those that express either resultative or progressive, or only resultative.

The examination of lexical homogeneity and *te-iru* thus explicated the fundamental concept of aspect—occurrence of "change" in the subject. Although *te-iru* functioning as an energy operator locates pure stative predicates such as adjectives in a secondary class of aspect, examination of lexical homogeneity results in a broadening of aspectual categories in Japanese, including pure stative predicates.

22. Some stative verbs such as *wakaru* "to understand," and *dekiru* "can do," however, can take *te-iru* when the resultative aspect is emphasized. Also, the existence verb *iru* that is used to describe the existence of animate things can be viewed as a volitional activity expressing the meaning of "to stay." *Iru*, however does not allow attachment of *te-iru* in standard Japanese.

4.4 NEGATION AS AN ASPECTUAL CATEGORY

The Japanese negative suffix *nai* is considered an adjective due to its morphological identicality to other adjectives. *Nai* can also be treated as an adjective aspectually, since adjectives are homogeneous; therefore, the assumption is that the negative suffix *nai* has the effect of converting heterogeneous predicates to homogeneous predicates. Please note the following examples:

(19) a. Saikin kyuuni futot-ta.
recentlysuddenly gain weight-PST
"Recently I suddenly gained weight."

b. Sorekara hantoshikan wa futor-anakat-ta.
thereafter half a year CNT gain weight-NEG-PST
"Since then I have not gained any weight for six months."

(20) a. Sono shunkan ni denki ga kie-ta.
that moment at light NOM turn off-PST
"The light was turned off instantly."

b. Roosoku no hi wa nagai aida kie-nakat-ta.
candle GEN light TOP long time turn off-NEG-PST
"The candle flame was not put out for a long time."

When the heterogeneous predicates *futoru* "to gain weight" and *kieru* "to turn off" are converted to negative forms, they express an unchanging situation with the ability to co-occur with adverbial phrases expressing durativity. This means that the aspectual character of the corresponding affirmative form of heterogeneous predicates no longer holds in the negative form. To negate a situation or an event is to portray nonexistence or nonoccurrence of the situation or the event. A negated event is a static state, not a dynamic event, and it would appear to qualify as a purely homogeneous phenomenon. Nevertheless, there are some features of negation that disprove this assumption. *Nai,* in some cases, has features similar to its corresponding affirmative form. Similar arguments were first put forth by

Tokieda (1955), who claimed that *nai* vacillates between objectified notions and subjectified notions. According to Tokieda, *nai* can independently constitute a proposition, just like an adjective, and can be dependently attached to a predicate. His discussion, however, was limited to the semantic behavior of *nai* and was largely based on his own intuitive judgment.

4.4.1 *Nai* and *Te-iru*

It is worth investigating Tokieda's argument in order to clarify the aspectual features of the negative suffix *nai*. The following examples are cases where *te-iru* variously can or cannot be attached to the negative suffix *nai:*

(21)	a.	Karada	ni	warui	kara	tabako	wa	suw-anaide-iru.
		health	for	bad	and so	cigarette	CNT	smoke-NEG-PROG

"Since cigarettes are bad for my health, I'm trying not to smoke."

	b.	Mainichi	tabako	o	takusan	sutte-iru.
		everyday	cigarette	ACC	many	smoke-PROG

"I'm smoking many cigarettes everyday."

(22)	a.	Kodomo	ga	jitto	nak-anaide-iru.
		child	NOM	patiently	cry-NEG-PROG

"The child is patiently trying not to cry."

	b.	Kodomo	ga	zutto	naite-iru.
		child	NOM	long	time cry-PROG

"The child has been crying for a long time."

(23)	* a.	Sakki	kara	doa	ga	ak-anaide-iru.
		while ago	since	door	NOM	open-NEG-PROG

"The door has not tried to open since a little while ago."

b. Sakki kara doa ga aite-iru.
while ago since door NOM open-RES
"The door has been open since a little while ago."

Example sentences (21)–(23) are cases where *te-iru* variously can or cannot be attached to the negative suffix *nai. Te-iru* can be attached to the negative form of the transitive verb *suw-anai* in sentence (21)a and the intransitive verb *nak-anai* in sentence (22)a. However, the negative form of the intransitive verb *ak-anai* in sentence (23)a does not allow the attachment of *te-iru.* The *suw-anaide-iru* in sentence (21)a is interpreted as the situation "not to smoke" and is maintained by the agent in the same fashion as the activity *sutte-iru* in (21)b except that *suw-anaide-iru* requires an inhibitional energy that is constantly controlled by the agent. This inhibitional energy is concealed in the agent and is balanced by the actual energy concurrently being input to the event. Neither *sutte-iru* nor *suw-anaide-iru* exists without the agent's constant input of energy, and the intransitive verb *nak-anaide-iru* in (22)a can be interpreted in the same manner. Maintained by the agent's intention, the effort "not to cry" requires energy in the same way as the antonymous activity, "to cry."

Based on the evidence that *nai* can co-occur with *te-iru* when an event expressed in the predicate can be affected by the agent's volitional control, the situation expressed by *nai* can receive energy in the same way as the original verb before being converted to the negative form.

These examples show that *nai* can, at times, function more like the affirmative form of activity verbs than pure stative forms, since pure states do not require constant energy input in order to maintain themselves. One difference, however, is that once an activity has begun, that activity can be considered to have a kind of inertia that works to keep the activity going and does not require much energy to maintain its dynamic situation. Inhibitive energy does not actually cause a change in a subject but acts as a kind of invisible counterweight to maintain a negative situation. This energy is static, but once the situation has been overwhelmed, it is radically altered as the agent bursts into a previously inhibited activity. The *naide-iru* form in some cases has the ability to contain stronger energy than is produced to generate the actual progressive activity. This hypothesis can also be examined by observing *nai* in the embedded clause *nai uchi ni*.

4.4.2 *Nai* and *Te-iru* in the *Uchi ni* Construction

The subordinate clause pattern *uchi ni*, consisting of the particle *ni* and *uchi* "within"[23] or "inside" as a whole means "while (i.e., while something is in a particular state)" or "before (before something happens)." The use of *uchi ni* is illustrated in example sentences (24) and (25):

(24) * a.

Densha	ga	tsuku	uchi ni	shinbun	o	kat-ta.
train	NOM	arrive	while	newspaper	ACC	buy-PST

"I bought a newspaper while the train arrived."

b.

Densha	ga	tsuk-anai	uchi ni	shinbun	o	kat-ta.
train	NOM	arrive-NEG	before	newspaper	ACC	buy-PST

"Before the train arrived I bought a newspaper."

(25) * a.

Kodomo	ga	oyogu	uchi ni	puuru	no	mizu	o	torikae-ta.
children	NOM	swim	while	pool	GEN	water	ACC	change-PST

"We changed the water of the pool while the children swam."

b.

Kodomo	ga	oyog-anai	uchi ni	puuru	no	mizu	o	kae-ta.
children	NOM	swim-NEG	while	pool	GEN	water	ACC	change-PST

"Before the children swam we changed the water of the pool."

23. *Uchi* can take either the particle *ni* or *wa* with varying consequences. The main sentence of an *uchi wa* construction tends to include a stative predicate, while the main sentence of an *uchi ni* construction tends to include a nonstative predicate as indicated in the following examples.

(a)

Tonari	no	ko	ga	ko-nai	uchi wa	shizuka-dat-ta.
neighbor	GEN	child	NOM	come-NEG	before	quiet-COP-PAST

"Before the neighbor's child came it was quiet."

While the heterogeneous predicate *tsuku* in sentence (24)a and the homogeneous predicate *oyogu* in (25)a are incompatible with *uchi ni,* their negative forms, *tsuk-anai* and *oyog-anai* are compatible with it. *Tsuk-anai uchi ni* in (24)b literally indicates the situation "while the train is in the state of not arriving." Nevertheless, as the English equivalents indicate, *uchi ni* is generally interpreted three different ways depending upon the predicate that occurs with it. The negative suffix *nai* co-occurring with *uchi ni* emphasizes "before something happens, or before the situation changes." *Te-iru* and an adjective co-occurring with *uchi ni* emphasize "during the period that something is in a particular situation." When a nominal co-occurs with *uchi ni,* the *uchin* clause becomes an adverbial phrase with a meaning such as "during a certain period of time." These sentences indicate that the speaker expects a change in the situation, and they imply a sense of urgency. There is, however, a difference in these examples in terms of the degree of urgency. Example sentences (26) and (27) illustrate this point.:

(26) a. Ame ga fur-anai uchi ni sentakumono o ire-yoo.
rain NOM fall-NEG before laundry ACC take in-VOL
"Before it rains I should take the laundry in."

b. Harete-iru uchi ni sentakumono o ire-yoo.
sunny-PROG while laundry ACC take in-VOL
"While the weather is sunny I should take the laundry in."

(27) a. Kuraku-nar-anai uchi ni uchi e kaet-ta.
dark-become-NEG before home LOC return-PST
"Before it got dark I returned home."

(b) Tonari no ko ga ko-nai uchini shigoto o shite-shima-oo.
neighbor GEN child NOM come-NEG before work ACC do-AUX -VOL
"Before the neighbor's child comes I should finish my work."

For the purposes of this study, only the *uchini* construction is considered.

b. Akarui	uchi ni	uchi	e	kaet-ta.
light	while	home	LOC	return-PST

"While it was light I returned home."

In these sentences, the negative and affirmative forms in the subordinate clauses (a and b, respectively) communicate the same result. Nevertheless, there is a psychological motivation for choosing one expression over the other. While the situation in sentence (26)b "to be sunny" is already realized and is being maintained, (26)a does not express a realized situation, but implies what a situation is like before a certain event is realized. This is why the sentence is interpreted as "before." This "before" means that the situation has not yet entered a different state, but it is very close to entering it. Both (26)a and (27)a imply that, at any moment, the given situations will change into the situations expressed in the affirmative forms, *ame ga furu,* and *kuraku-naru,* respectively. These sentences are often used to suggest some kind of warning in preparation for a change in a situation. The content of the change is clearly indicated in the corresponding affirmative form. On the contrary, sentences (26)b and (27)b do not imply that any major change is expected, especially in comparison to a negative form combined with *uchi ni. Harete-iru uchi ni* in (26)b implies, with a sense of urgency, that the situation will change from a sunny to a nonsunny weather pattern. This could be any type of weather: cloudy, snowy, windy, or rainy—as long as it is not sunny.

It is worth noting that the opposite of a given negative form is generally considered its affirmative, but the opposite of a given affirmative is not necessarily the corresponding negative. An affirmative form is certainly more likely to have a greater number of viable opposite considerations. *Harete-iru uchi ni* is not as clear as *fur-anai uchi ni* in terms of what kind of change is brought to a situation. Where *harete-iru uchi ni* implies some change, but not any particular change, *fur-anai uchi ni* additionally implies a particular result, in this case that the situation will change from not raining to raining.

As Dowty (1979) notes, states and activities share the property of being homogeneous. A similar observation can also be seen in Kindaichi's argument concerning the definition of verb categories. Kindaichi (1976b) stated that "continuous verbs are verbs which indicate a temporary change of state" (p. 24). This is to say that in a sentence containing an activity predicate, the subject itself does not change, but the performance or nonperformance of the activity by the subject changes. This, however, is not how stative predicates are viewed. Stative predicates are rather closely related to the notion of "everlasting," but activities are probably more receptive to change after lasting a certain period of time because of the energy levels needed to maintain them. A person can eat, walk, and swim, for example for only a certain period of time. It is generally assumed that the negative suffix *nai* is a stative construction and as such should adhere to this "everlasting" notion. However, *nai* in the *uchi ni* construction acts more like an activity predicate than a stative predicate. The predicate *oyog-anai* in sentence (25)b is more closely related to the affirmative form *oyogu* "to swim" than it is to real stativity, because the situation expressed in *oyog-anai* can be changed as easily as the situation expressed in *oyogu.* The fact that *nai uchi ni* relays a strong sense of urgency can be compared to another embedded clause with *mae ni,* which literally means "before." Please note the following examples:

(28)	Kodomo	ga	oki-nai	uchi ni	kaimono	o	shite-shima-oo.
	child	NOM	wake up -NEG	before	shopping	ACC	do-AUX-VOL

"I should go and finish shopping before the child wakes up."

(29)	Kodomo	ga	okiru	mae ni	kaimono	o	shite-shima-oo.
	child	NOM	wake up	before	shopping	ACC	do-AUX-VOL

"I should go and finish shopping before the child wakes up."

Both sentences have the same English equivalent, although there is the syntactic difference that *mae ni* takes only the affirmative form of a predicate while for *uchi ni* this is not the case. *Mae ni* indicates a concrete point of reference for *okiru* "to wake up," and the speaker is referring to the time the event is realized. *Nai uchi ni,* however, cannot reference the point at which the event is realized because negation itself denies the realization of the event. *Nai uchi ni* is an assemblage of an iterated, nonrealized event that comprises a time expanse. With the use of *uchi ni*, the possibility that an event will be realized suddenly, at any particular instant, is implied. This communicates with a sense of urgency that the situation will change. *Mae ni*, however, does not create exactly the same implication since it points to the concrete time reference expressed by the affirmative form.

When the *uchi ni* construction is used, the main clause plays a large role in deciding the meaning of the whole sentence. The examples we have seen so far have volitional forms and past events in the main clauses. Another interpretation is shown in sentence (30):

(30)	Hanashite-iru	uchi ni	dandan	kanojo	ga	suki ni-natte kita.
	talk-PROG	while	gradually	she	NOM	like-become-PAST

"I came to like her while I was talking with her."

This example communicates the fact that the event in the main clause was realized during the time of the activity indicated in the subordinate clause. In sentence (30), the speaker's feeling changed gradually while the speaker was talking with someone. Such cases are typically expressed using *kuru* or *naru* in the main clause, indicating a nonwillful change within the area from one endpoint to the other endpoint of an interval. It expresses that the event in progress parallels the event in the embedded clause. This subordinate clause implies neither a change in the situation expressed in the subordinate clause nor a sense of urgency. The activity is in a state of inertia until the event in the main clause reaches the other end of an interval. In this case, the activity occurs between two points of an interval and this interval can be expressed using *aida ni*.

Uchi ni can often be rephrased *aida ni*, "while something is in the state of ~." Aida *ni* literally means "while, or between two points of time or space" and expresses a duration of time or distance between two objects. *Uchi ni* can often be replaced by *aida ni* since both co-occur only with homogeneous predicates, as in the following examples:

(31) Chotto aw-anai uchi ni/aida ni ookiku-nat-ta ne!
a little while see-NEG while big-become-PST PART
"While I was not seeing you, you grew a lot, didn't you!"

(32) Shir-anai uchi ni/aida ni ha ga waruku-natte-i-ta.
know-NEG while tooth NOM bad-become-RES-PST
"While I was not aware, my teeth had gotten bad."

In example sentences (31) and (32), the predicates of each main clause, which indicate change within an interval, give rise to the interpretation that *nai uchi ni* expresses a simultaneous progression between the events in the main and subordinate clauses. Both *aw-anai* and *shir-anai* define an interval between two endpoints, and the sentences are not interpreted as "before," but support a "while" interpretation. These *nai uchi ni* constructions do not create awkwardness when interchanged with *aida ni*; however, when the predicate in the main clause does not indicate an interval, but rather a single time event, this is no longer the case:

(33) a. Kodomo ga oki-nai uchi ni kaimono o shi-yoo.
child NOM wake-NEG before shopping ACC do-VOL
"Before the child wakes up I should go shopping."?

b. Kodomo ga oki-nai aida ni kaimono o shi-yoo.
child NOM wake-NEG while shopping ACC do-VOL
"While the child does not wake up I should go shopping."

c. Kodomo ga nete-iru aida ni kaimono o shi-yoo.
child NOM sleep-PROG while shopping ACC do-VOL
"While the child is sleeping I should go shopping."

When *aida ni* is used instead of *uchi ni* in sentence (33)b it is somewhat awkward, although it should be noted that this type of sentence occurs occasionally in colloquial speech without drawing much attention to itself. Sentence (33)c, however, is more appropriate in the same situation. This is because the word *aida* points toward a duration of time between two endpoints of an interval (*aida*'s literal meaning). Activities such as *neru* "to sleep," clearly indicate inception and termination of an activity, and while the activity occurs between two points in time, it maintains progression through the input of iterative energy.

In a resultative situation, energy is not infused in the same manner as in an activity, but rather attains a state of inertia in which human volitional energy input does not play a part. This can be seen in the following example with an intransitive verb in the subordinate clause:

(34)	Doa	ga	aite-iru	uchi ni	neko	ga	hairikonde-ki-ta.
	door	NOM	open-PERF	while	cat	NOM	enter-PST

"While the door was open a cat came in."

This sentence literally expresses that a cat entered while the door was open, and implies that the cat came in before the door was shut. This use of *uchi ni* implies that a change would occur in the situation and implies a sense of urgency. This change can be precipitated by either the force of human volition or some natural force. Therefore, when a situation can no longer be changed by any kind of force, *uchi ni* is an inappropriate choice. Please note the following examples:

(35)*	Inu	ga	shinde-iru	uchi ni	neko	o	asob-ase-yoo.
	dog	NOM	die-PERF	while	cat	ACC	play-CAUS-VOL

"While (during the time) the dog is dead I should let the cat out to play."

(36)?	Hikooki	ga	tsuite-iru	uchi ni	omiyage	o	kat-ta.
	airplane	NOM	arrive-PERF	while	souvenir	ACC	buy-PST

"While the airplane was arriving I bought some souvenirs."

In these examples, none of the verbs accompanying *uchi ni* has the potential to recover from their original states: "dead" to "not dead" or from "arrived" to "not arrived." Although *uchi ni* implies that some kind of change occurs to the state, the state must be one that inherently possesses the possibility of change due to human or natural forces.

What was observed through the above *uchi ni* constructions is that *te-iru uchi ni* expresses two different meanings, depending upon the main clause. The first is that during a situation described by the subordinate clause, a one-time event described in the main clause occurs, as indicated in sentence (37):

(37) Kodomo ga soto de asonde-iru uchi ni sooji o shi-ta.
child NOM outside LOC play-PROG while clean ACC do-PST
"I cleaned (the room) while my child was playing outside."

The second meaning expressed is that within a passage of time occurring in a situation described by the subordinate clause. A change expressed in the main clause is taking place in temporal concert with the interval expressed in the subordinate clause, as seen in sentence (38):

(38) Hon o yonde-iru uchi ni nemuku-nat-ta.
book ACC read-PROG while sleepy-become-PST
"While I was reading a book I became sleepy."

In both cases, "while (during)" is the approximate equivalent in English.

Nai uchi ni also expresses two meanings that differ depending upon the main clause. One is that before a change occurs in a situation described by the subordinate clause, the one-time event described by the main clause occurs. This was shown in sentence (33)a *Kodomo ga oki-nai uchi ni kaimono o shi-yoo.* The second is the same as the second meaning of *te-iru uchi ni:* a change takes place in the main clause which is in temporal concert with an interval of time defined by the subordinate clause. This was shown in sentence (31) *Chotto aw-anai uchi ni ookiku-nat-ta ne.*

This analysis shows that there is common ground between *te-iru* and *nai* in that both can co-occur with a deliberate change occurrence expressed by main clause predicates such as *naru, kuru* and so forth. *Chotto aw-anai uchi ni ookiku-nat-ta ne* in example (31) shows that the nonoccurrence of the event *aw-anai* "not to see" develops in parallel to the occurrence of the event in the main clause, *ookiku-naru.* The occurrence of the event entices the nonoccurrence of the event into its time frame and marks the starting point and ending point of an interval. Activity predicates such as *yomu* and *asobu* can express these two end points of an interval as inception and termination of the activity. The negative suffix *nai* generally does not express two such ends of an interval, since *nai* expresses unrealized, nonoccurring, and static situations in which we cannot clearly observe any such end-points. However, *nai* can also convey a similar situation to that of *te-iru,* one in which the interval is limited to a recognizable duration of time in the actual world. The only difference is that *nai* lacks the visual dynamics of an activity itself.

In contrast, the different features of *te-iru uchi ni* and *nai uchi ni* are shown in their interpretations "while" and "before." Although both imply the occurrence of a change, they exhibit a difference in terms of the degree of the sense of urgency, that is, the possibility of change indicated in the *uchi ni* construction. For example, *nete-iru uchi ni* expresses that the activity "to sleep" has already entered into a progressive state. The existence of this progressive state indicates that it is a situation that has begun at some nonspecified time, and also implies that the subject will eventually arrive at the point of waking up, the other endpoint defining the sleep interval. The activity "to sleep" is a kind of state of inertia occurring in between these two points of the interval. *Nai uchi ni,* on the other hand, does not point to a clear starting point of an activity; rather, it only indicates a point that connects a nonactual world to the actual world—a static world to a dynamic world.

The features of *nai* and *te-iru* were examined by observing their behavior in the *uchi ni* construction, and in cases where *nai* co-occurred with *te-iru.* Although it was natural to assume that the negative suffix *nai* has the effect of converting nonstative predicates to stative predicates, it was shown that *nai* at times behaves in ways similar to the original affirmative form of the predicate. Unlike other purely stative predicates, such as adjectives and the copula *da, nai* can indicate a situation where energy is applied, as was seen in sentences that had *nai* co-occurring with *te-iru.* This energy is mainly input by a subject that possesses volitional control over a situation. *Nak-anaide-iru,* for example, requires a larger amount of maintenance energy compared to a situation in which someone is crying. Whenever some kind of change is implied to occur, energy is concealed until the moment the change is brought about. This was also seen in the use of the *uchi ni* construction. *Nai-uchi ni* conceals the highest potential energy, that is, the highest degree of certainty that a situation will change. This means that *nai* can possess, at times, a large capacity to store energy, at least in comparison to *te-iru.*

It is interesting to note that the characteristics of *nai* and *te-iru* in the *uchi ni* construction are analogous to a closed system of potential/kinetic energy in elementary physics. For example, at the moment a pendulum reaches its apex the energy of its movement is 100% potential and 0% kinetic. That is, the pendulum is in a static state (at rest), but completely imbued with potential energy. As it swings toward the vertical position the energy pattern shifts. Along the way it will at various times be 80% potential and 20% kinetic, then 60%/40%, and eventually, when the pendulum is completely vertical, at its lowest point and moving its fastest, the energy in the system can be expressed as 0% potential and 100% kinetic. The trend then reverses itself until the pendulum arm reaches the opposite apex. In this analogy *nai* can be thought of as potential energy and *te-iru* as kinetic energy. *Nai* situations tend to be static, but have the highest potential for change. This was evident in the fact that *nai uchi ni* conveys that there is a high possibility that a situation is going to change at any time. *Te-iru uchi ni* on the other hand communicates the fact that some kind of event is occurring at a particular moment. Although it is not possible, or necessarily desirable here, to force a parallel between language phenomena and physics to the extent that we would find a sentence expressing a state that is 80% potential and 20% kinetic, for example, the description of situations using aspectual forms does exhibit a correlation with basic physics. The following sentences, taken from previous examples, represent this potential/kinetic relationship:

(33)a Kodomo ga oki-nai uchi ni kaimono o shi-yoo.
child NOM wake-NEG before shopping ACC do-VOL
"Before the child wakes up I should go shopping."?

(37) Kodomo ga soto de asonde-iru uchi ni sooji o shi-ta.
child NOM outside LOC play-PROG while clean ACC do-PST
"While the child was playing outside, I cleaned the room."

(38) Hon o yonde-iru uchi ni nemuku-nat-ta.
book ACC read-PROG while sleepy-become-PST
"While I was reading a book I became sleepy."

Sentence (33)a represents a 100% potential and 0% kinetic situation, (38) a 0% potential and 100% kinetic situation, and (37) lies somewhere in between.

As illustrated in Figure (39)a below, the pendulum at point A (representing the *nai uchi ni* construction) has maximum potential energy, has the maximum potential for change, and carries with it the highest sense of urgency. Points B and C (representing *te iru uchi ni* constructions) have reduced potential energy and a smaller sense of urgency. The plot in Figure (39)b shows the potential energy, U (associated with the potential for sudden change and, hence, the sense of urgency), and the kinetic energy, K, for the pendulum system plotted for each case:

(39)a

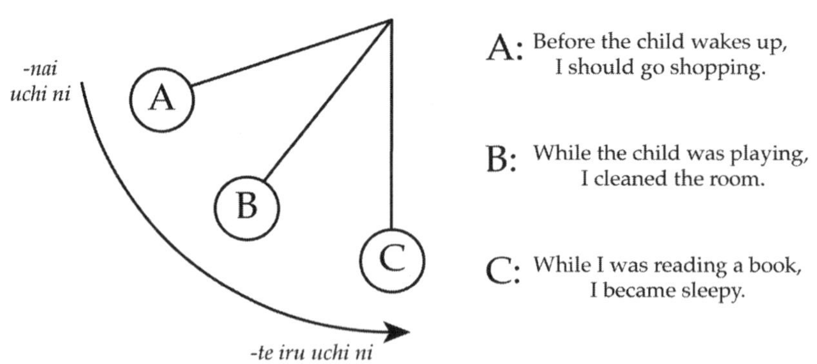

(39)b

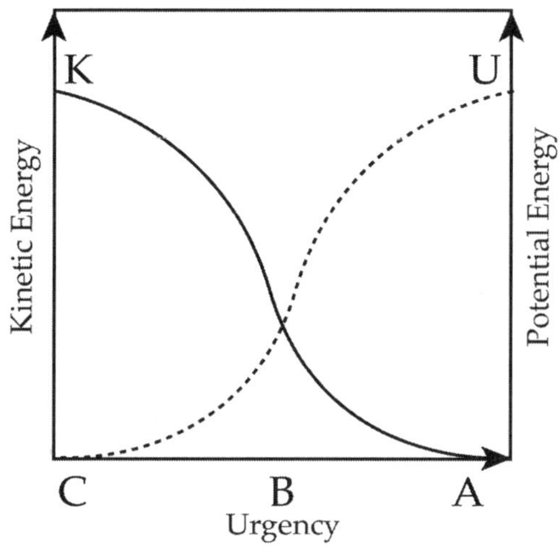

4.4.3 Summary of *Nai* and *Te-iru* as Aspectual Categories

The features of *nai* and *te-iru* were examined by observing their behavior in the *uchi ni* construction and in cases where *nai* co-occurred with *te-iru*. Although it was natural to assume that the negative suffix *nai* has the effect of converting heterogeneous predicates to homogeneous predicates, it was shown that *nai* at times behaves in ways similar to the original verb affirmative form. Unlike other purely homogeneous predicates, such as adjectives and the copula *da*, *nai* can indicate a situation where energy is applied, as was seen in sentences in which *nai* co-occurred with *te-iru*. This energy is mainly input by a subject that possesses volitional control over a situation. *Nak-anaide-iru* "remain in the state of not crying," for example, requires a larger amount of energy compared to a situation where someone is crying. Whenever some kind of change is implied to occur, energy is concealed until the moment the change is brought about. This was also seen in the use of the *uchi ni* construction. *Nai-uchi ni* conceals the highest potential energy, that is the highest degree of certainty that a situation will change. This means that *nai* can possess, at times, a large capacity to store energy, at least in comparison to *te-iru*.

Nai and *te-iru* thus often behave similarly, especially in terms of aspect, vacillating between a dynamic world and a static world depending on the environment in which they are used. This semantic vacillation can also be seen in their syntactic behavior, namely in terms of internal negation and external negation. In the next section, further examination of these structures will be given in terms of modality.

4.5 THE LIMITS WITHIN WHICH NEGATION EXPRESSES MODALITY

In this book, modality is defined in part as expressing possible/nonactual worlds. Possible worlds are those concerned with the way things may be or might have been as viewed by the speaker at the time of speech. Present events that do not invite the speaker's conjecture, habitual events, and events accepted as general fact or scientific truth, for example, are contrarily in the domain of the actual world. For instance, the sentence *okane ga aru*, "I have money," is a description of the speaker's situation at the time of speech and is in the actual world domain. If the situation is *okane ga nai*, "I do not have money," the fact that the speaker does not have money is considered to fall in the actual world domain. In the latter example, having money is a nonactual world situation. Thus, actual worlds can be described by either affirmative or negative sentences, depending on the situation a speaker is in. Consequently, when negative sentences describe actual worlds, they do not express modality. On the other hand, when negative sentences describe nonactual world situations, a correlation between negation and modality can be detected in the ability of negation to express nonactual world situations. Hence, negation appears to fall under the domain of modality. The following negative sentences explore the possibility of expressing modality in actual and nonactual world situations:

(40) Ano eiga, omoshiroku-nai yo.
that movie interesting-NEG VOC
"That movie is not interesting, I'm telling you."

(41) a. Otto wa asagohan o tabe-nai.
husband TOP breakfast ACC eat-NEG
"My husband does not eat breakfast."

b. Otto wa asagohan wa tabe-nai.
husband TOP breakfast CNT eat-NEG
"My husband does not eat breakfast, (but implies that he eats lunch, for example)."

(42) Musuko wa basu de ik-anakat-ta.
son TOP bus INS go-NEG-PST
"My son did not go by bus."

Sentence (40) describes a movie's attributes, namely that a particular movie is not interesting, with the adjective "interesting" serving as the focus of negation. This negative sentence therefore expresses an actual situation. Sentence (41)a describes a husband's habitual noneating of breakfast. The verb *taberu* "to eat" is the focus of negation, and *tabe-nai* "does not eat" is a fact that falls in the actual world domain. Sentence (41)b, however, does not receive the same interpretation as (41)a due to the use of the particle *wa* replacing the object marker *o;* it receives a contrastive meaning implying that the husband does not eat breakfast, but eats lunch and/or dinner. This is a standard view adopted by most linguists.

McGloin (1976a), for example, using a model of generative semantics, introduced the contrastive *wa* in terms of the interaction of presupposition and assertion of a sentence, and claimed that "the contrastive *wa* marks the scope of negation" (pp. 390–92). As McGloin suggests, contrastive *wa* plays a significant role in determining the scope of negation as seen in example sentence (41)b, *Otto wa asagohan wa tabe-nai* "My husband does not eat breakfast (but implies that he eats lunch, for example)."

Kuno (personal communication 2001), however, asserts that there is no reason why sentence (41)b should be interpreted that way; it can be interpreted as meaning "Speaking of breakfast, as opposed to lunch and dinner, my husband does not eat breakfast," in which the focus of negation is the verb *taberu.* This is exactly how the focus of negation is perceived for the sentence *Taroo wa ko-nakat-ta* "Taroo did not come." For this sentence, the noncontrastive and contrastive interpretations of Taroo are ambiguous although most linguists would assume that the focus of negation is Taroo. Kuno also asserts that there is another way to account for the contrastive meaning of the *wa*-marked subject. That is, what the sentence means is "Speaking of Taroo, and not of the rest, he did not come." This is the equivalent of "Limiting my talk to Taro, he did not come," in which the focus of negation is on the verb *kuru* "to come."

Therefore, noncontrastive and contrastive interpretations of a sentence are usually context-dependent. Kato (1985) similarly states that the topic-contrastive distinction is discourse-dependent, and the distinction is a nonbinary opposition in that a single occurrence of *wa* can be a topic and a contrastive simultaneously. Sentence (41)b can therefore be interpreted not only as "Speaking of my husband, he does not eat breakfast (implies that he eats lunch and/or dinner)," but also as "In contrast with my brother, for example, my husband does not eat breakfast (implies that he eats lunch and/or dinner)," with the latter indicating a case of mutual contrast.

Whatever the case may be, the contrastive *wa* is used when a speaker tries to imply a situation that is somehow compared to that which is uttered. To use a contrastive structure is to focus on the meaning behind what is spoken. Therefore, when contrastive *wa* is used, it cues the interlocutor to the distinguished properties of what the speaker is actually articulating, that is, there is a presence of an "invited inference."[24] The speaker is conveying that the husband actually eats lunch and/or dinner, in which the eating activity is considered in the actual world domain. Therefore, *tabe-nai* in (41)b can refer to a nonactual world situation.

Although McGloin's premise (i.e., that contrastive *wa* defines the scope of negation) is true to a great extent, as she later pointed out (1990), it should also be recognized that *wa* is not the only element that helps define that scope. Example sentence (42*), Musuko wa basu de ik-anakat-ta,* "My son did not go by bus," is evidence for a case where *wa* is not used in this way. A question arises, however, as to why there are such cases where a constituent preceding a negative predicate can become the focus of negation even without the appearance of the contrastive *wa*.

Sentence (42) does not involve the contrastive marker *wa*. The focus of negation, however, is not on the verb *iku*, "to go," and this sentence carries the implied meaning that the son actually did go, not

24. Refer to chapter 3, section 3.2.4, where *kamoshire-nai* is discussed.

by bus, but by some other means of transportation. *Ik-anakat-ta* "did not go" therefore refers to a nonactual world situation. Although the absence of any device can let us envision an implied situation, there are two ways to further explicate such meaning: the use of contrastive *wa* after the phrase *basu de* "by bus" and the use of external negation structure, *musuko wa basu de it-ta-no dewa-nai*, "it is not that my son went by bus." The external negation structure is formed by using the negative form of the expressions *no-da* or *wak-da*, which are also considered modals (See chapter 2, figure [14]).[25] The use of these modals suggests a strong correlation between modality and negation in terms of possible/nonactual world situations. It is now possible to investigate the circumstances that cause this to occur, and further to examine whether or not these situations are indeed in the domain of modality as it is outlined herein.

4.5.1 The Scope of Negation and Modality: Internal Negation[26]

On the basis of Reinhart's argument (1983), the scope of negation is defined as the syntactic domain of the negative morpheme, and the focus of the negation must lie within the scope of negation. A similar definition was proposed by Sagawa (1976): the term "scope" refers to "the total sum of the domains where the focus of negation can appear." The scope of negation in Japanese is most notably examined by Kuno (1983), who proposed a definition of scope based on a discourse-oriented approach that ultimately influenced the semantic theory of negation in Japanese. According to Kuno, the scope of negation does not extend over the constituent to which the negative morpheme *nai* is attached. It only affects predicates that immediately precede it. There are, however, cases where the syntactic focus of negation does not agree with the semantic focus of

25. Although Ohta (1980) states that the distinction between internal negation and external negation is not related to truth conditions, but rather to the role they play in discourse, in this book these terms are used to refer to structural differences. External negation is restrictively expressed by *(to iu) wake de wa nai, n(o) de wa nai,* for example. This is a type of negation that is equivalent to English expressions such as "it is not the case that . . ." or "it is not that . . ."

26. Internal negation is a structure in which the negative element negates only a part of a sentence. This is opposed to external negation structures in which the negative element negates the entire sentence.

negation.[27] Kuno (1983, p. 130) approached this problem by raising two sets of exceptions: 1.) when the underlying constituents in question are selected from a closed set of alternatives (multiple choice); and 2.) when the constituents in question are selected from an open set (fill-in-the blank). Multiple choice is applicable when the type of element that becomes the focus of negation is fairly limited and the type of event may be repeatable (such as "buy a watch"), while fill-in-the blank is applicable when the type of element that becomes the focus of negation is not limited (open-set) and usually the type of event is a one-time event (such as "to be born"). Although Kuno claimed that these notions of multiple choice and fill-in-the blank are dependent on context, his theory implies that the number of elements that can be the focus of negation is inherent in the constituent before it is negated. A question arises, however, as to why cases like multiple choice and fill-in-the blank need to be established in Japanese. Is it attributed only to the structural characteristics of Japanese that the negative morpheme *nai* immediately follows a predicate?

Approaching this problem via predicate type may help to shed some light on the focus of negation. Every predicate has a restriction on the range of categories that a given item permits or requires

27. Kuno gave the following examples (1983, p. 126):

(a) A: Kimi wa shuusen no toshi ni wa
you TOP end of the war NOM year TEMP CNT
moo umarete-i-ta-no ka.
already born- PERF-PAST-COMP Q
"Had you already been born in the year when the world war was over?"

B: Iya, shuusen no toshi ni wa
no end of the war NOM year TEMP CNT
mada umarete-i-nakat-ta.
yet born-PERF-NEG-PAST
"No, I was not yet born that year."

(b) A: Kimi wa kono shashin o Pari de tot-ta-no ka
you TOP this picture ACC Paris LOC take-PAST-COMP Q
"Did you take this picture in Paris?"

* B: Iya, Pari de (wa) tor-anakat-ta.
no Paris LOC (CNT) take-NEG-PAST
"No I didn't take the picture in Paris."

Kuno explains that the appropriateness of the answer to the question in (a) is attributable to the fact that the semantic focus of negation *umarete-i* precedes the negative morpheme *nakat-ta* and therefore that focus is in the domain of the scope of negation. The inappropriateness of the answer in sentence (b) is, on the other hand, attributed to the disagreement of syntactic focus of negation (*tora-* part only) with semantic focus of negation (Paris).

as its complement, referred to as "subcategorization restrictions."[28] For example, *nomu*, "to drink," is a two-place predicate taking an obligatory Nominal Phrase (=NP) complement (object), and a ditransitive verb such as *oshieru*, "to teach," is a three-place predicate taking two obligatory NP complements (direct object and indirect object) in addition to the subject. What we will examine here primarily involves extremely simple sentences, and we will look not only at the number of obligatory NP complements a predicate takes, but also the possible number of nonobligatory NP complements it can take. These possible NP complements are usually locative, temporal, or instrumental lexical items. For example, suppose that each symbol represents a NP complement for the pseudo sentence, *wxyz tabe-ta.* One simple example sentence may be, *Watashi (wa) hachi ji (ni) uchi(de) gohan(o) tabe-ta*, "I ate breakfast at 8 o'clock at home." The predicate *tabe-ta* "ate" allows temporal and locative NP complements in addition to the obligatory NP complements (subject and object). Also, since the negative suffix *nai* is one of the aspectual categories, the notion of the homogeneity/heterogeneity dichotomy should be integrated in the examination. Below, a brief analysis is given regarding how these elements behave when the sentence is converted to its negative.

4.5.1.1 Heterogeneous Predicates and the Focus of Negation

First, we examine heterogeneous predicates, using *kaeru*, "to return," *tsuku*, "to arrive," and *umareru*, "to be born" (the first involving a complex change of state, and the latter two involving a single change of state):

(43) a. Miwa wa kaer-anakat-ta.
Miwa TOP return-NEG-PST
"Miwa did not return."

b. Miwa wa Tookyoo e kaer-anakat-ta.
Miwa TOP Tokyo LOC return-NEG-PST
"Miwa did not return to Tokyo."

28. Radford 1988, p. 369.

c. Miwa wa densha de Tookyoo e kaer-anakat-ta.
Miwa TOP train INS Tokyo LOC return-NEG-PST
"Miwa did not return to Tokyo by train."

d. Miwa wa juuichi ji ni Tookyoo e kaer-anakat-ta.
Miwa TOP 11 o'clock TEMP Tokyo LOC return-NEG-PST
"Miwa did not return to Tokyo at 11 o'clock."

(44) a. Hikooki ga tsuk-anakat-ta.
airplane NOM arrive-NEG-PST
"The airplane did not arrive."

b. Hikooki ga kuukoo ni tsuk-anakat-ta.
airplane NOM airport LOC arrive-NEG-PST
"The airplane did not arrive at the airport."

c. Hikooki ga san ji ni kuukoo ni tsuk-anakat-ta.
airplane NOM 3 o'clock TEMP airport LOC arrive-NEG-PST
"The airplane did not arrive at the airport at 3 o'clock."

(45) a. Sono ko wa umare-nakat-ta.
that child TOP born-NEG-PST
"That child was not born (The child died prior to birth)."

b. Sono ko wa yooka ni umare-nakat-ta.
that child TOP 8th of the month TEMP born-NEG-PST
"That child was not born on the eighth of the month."

c. Sono ko wa Tookyoo de umare-nakat-ta.
that child TOP Tokyo LOC born-NEG-PST
"That child was not born in Tokyo in January."

The heterogeneous predicates *kaeru*, "to return," and *tsuku*, "to arrive," are motion verbs that indicate the movement of an entity from one place to another. In comparing the interpretation of sentences (43)a and b, however, without the presence of the contrastive *wa*, the locative NP complement *Tookyoo e* "to Tokyo" in (43)b does not create a drastically different interpretation from that received by (43)a. Sentence (43)b does not particularly imply that Miwa returned to some place other than Tokyo. The actual event is that which is expressed in negation. Contrary to this, the instrumental NP complement *densha de* "by train" in (43)c, and the temporal NP complement[29] *juuichi ji ni* "at 11 o'clock" in (43)d immediately let us perceive that they are the focus of negation, that is, they let us envision the actual events that Miwa returned to Tokyo, but not by the last train, and that Miwa arrived in (or left for) Tokyo, but not at 11 o'clock.[30]

In sentence (44)b, the locative NP complement *kuukoo ni* "at the airport" does not seem to serve to focus the negation. It is difficult to infer that the actual event is the airplane's arrival at some place other than the airport. On the other hand, the temporal NP complement san *ji ni* "at 3 o'clock" in (44)c clearly implies that the actual event of the airplane's arrival is at some time other than 3 o'clock. The temporal NP complements are thus the foci of negation for both *kaeru* "to return" and *tsuku* "to arrive," but locative NP complements do not have the same behavior. This suggests that the relevant restriction is that, for motion verbs, the locative NP complement must be connoted even if not expressed. The semantic property of such motion verbs has an influence on the interpretation of the focus of negation. The presence of geographical information with these motion verbs is fundamental even if it is not

29. The term "temporal" is used to refer to punctual time adverbial expressions, such as *sanji*, "three o'clock," *kyoo*, "today," *ichigatsu*, "January," and *kyonen*, "last year," and is contrasted with the notion of "durative," which refers to a continuous interval in time.

30. Sentence (43)d, however, is in fact a bit awkward since the verb *kaer-anakat-ta* "did not return" and the punctual NP complement *juuichiji ni* "at 11 o'clock" do not mesh well. This is attributed to the feature of motion verbs that they involve geographical distance and require a certain time frame to accomplish the event, specifically an interval between the point of departure and the point of arrival. *Juuichiji ni*, "at 11 o'clock," does not indicate at which point the event "to return" is realized.

explicitly mentioned, since such information is understood from the discourse.[31]

Sentence (45)a describes an actual situation that the child was not born. Sentence (45)b, including a temporal NP complement is, again, very awkward and unstable. The temporal NP complement makes an interlocutor immediately draw the focus of negation, and subsequently the sentence alludes to nothing but the implied meaning that the child was born, but not on the eighth of the month. *Umare-nakat-ta* "was not born," therefore, can easily refer to a nonactual event in this interpretation. In sentence (45)c, unlike in sentences involving motion verbs, the locative NP complement *Tookyoo de* "in Tokyo" accompanying an instantaneous verb also draws the focus of negation; it creates the implication that the actual event is the birth of the child, but that the birth occurs at a place other than Tokyo. This observation suggests that locative NP complements may not serve to focus the negation for verbs that imply direction inherently, but for verbs that describe an event that takes place within a delimited geographical area.[32]

In the heterogeneous predicate category, it can be seen that, irrespective of the semantic properties of predicates, only temporal NP complements become the focus of negation. This can be understood from the fact that the notion of change in heterogeneous predicates is associated with the concept of time. When a heterogeneous predicate is involved in a sentence, there is a situational difference between

31. In addition to factors such as intonation and pragmatic context, word order may also determine the focus of negation. As Kuno (1983) states, the scope of negation does not extend over the constituent to which the negative morpheme *nai* is attached. This may suggest that the closer a complement is to the negative morpheme *nai,* the higher the chances are syntactically that the complement will be in the scope of negation, i.e., the farther apart a complement becomes from the negative predicate, the higher the chances that it becomes the focus of negation.

32. The particle *de* co-occurring with a locative complement functions to limit the range of the location/time frame in which an activity is performed or to limit the object with which the activity is performed. Therefore, it can co-occur with a locative NP complement or an instrumental NP complement. The former indicates a place where an activity takes place, and the latter indicates a method by which an activity is performed. With this basic function of *de* in mind, it is clear that *de* can also co-occur with various classifiers with numerals that limit the range of time in which an activity is performed, such as *in Kono shukudai wa mikka de shi-ta* "As for the homework, I did it (with)in three days." When this sentence is negated, *mikka* "three days" is the focus of negation. This suggests that the nonobligatory NP complements are usually the focus of negation, except for locative NP complements co-occurring with a verb with inherent direction. This postulation is reasonable since it is generally the case that mentioning an inessential element draws attention to that element.

"before" and "after." For example, with *shinu* "to die" the change is from alive to not alive, and with *futoru* "to become fat," the change is from not fat to fat. The notion of change with a time boundary (punctual or durative) is separated into two opposite situations. If such a temporal NP complement is accompanied by an affirmative predicate, it naturally alludes to the point where a change occurs, for example, as *in Sono ko wa yooka ni umare-ta* "That child was born on the eighth of the month." When an event is represented by a negative form, however, it is difficult to conceptualize the nonoccurrence of the event as an actual event occurring in a single instant. Therefore, when a temporal NP complement is accompanied by a negative heterogeneous predicate, the sentence is awkward to begin with. If not, there is a strong implication that the actual event corresponds to the affirmative form. Thus, the actual events in negative statements with temporal NP complements are mostly what can be described by the affirmative forms of the predicates. These negative forms of heterogeneous predicates co-occurring with temporal NP complements can express nonactual world events, yet the sentence needs to be converted with the use an external negation structure or with the use of the contrastive *wa* in order to avoid awkwardness.

4.5.1.2 Homogeneous Predicates (Activities) and the Focus of Negation

We turn our attention now to homogeneous predicates, considering both activity predicates and stative predicates:

(46) a. Toshokan de teepu o kik-anakat-ta.
library LOC tape ACC listen-NEG-PST
"I did not listen to the tape at the library."

b. Go ji ni teepu o kik-anakat-ta.
5 o'clock TEMP tape ACC listen-NEG-PST
"I did not listen to the tape at 5 o'clock."

As seen in the behavior of locative NP complement in sentence (45)c, locative NP complements tend to serve as the focus of negation, since geographical information may not have to be inherent for the activity predicate as much as in the case of verbs with inherent direction. The locative NP complement *toshokan de* "at the library" with an activity predicate in sentence (46)a thus creates an implication in which the actual event is "I listened to the tape at a place other than the library," and therefore serves as the focus of negation.[33]

Similar to the case of heterogeneous predicates, the temporal NP complement in sentence (46)b appears awkward. This is attributable to the blending of a punctual temporal NP complement *go ji ni* "at 5 o'clock" that indicates specific point in time at which the event commences with *kik-anakat-ta* "did not listen," which is not a punctual event. This type of event controlled by an agent's volition is viewed as being carried out within a certain time expanse. In daily conversation, the interpretations of these sentences are usually "I did not start the activity at such and such a time" or "I was not listening to the tape at five o'clock," which instead implies that a duration of time is involved in the activity. Therefore, the sentence is well received if a durative NP complement such as "from 4 to 5 o'clock" is supplied. The durative NP complements, however, do not function as strongly as foci of negation as punctual NP complements co-occurring with negative heterogeneous predicates. This is due to the semantic property of activity predicates that activities are volitionally controlled to continue or discontinue for a certain time frame, and the nonoccurrence of an event during a certain period of time is not difficult to conceptualize.

Another category of predicates in the homogeneous category is that of stative predicates, illustrated in the following examples involving an adjective *shiroi*, "white," and the existence verb *aru*, "(inanimate things) exist":

33. It should be mentioned that when the locative NP complements are specific in context, they definitely become the focus of negation. For example:

(a) Kyoo wa puuru de oyog-anakat-ta.
today TOP pool LOC swim-NEG-PAST'
"I did not swim in the pool today."

(b) Kyoo wa itsumo no puuru de oyog-anakat-ta.
today TOP regular GEN pool LOC swim-NEG-PAST
"I did not swim in the pool in which I regularly swim."

Sentence (b) clearly implies that the speaker swam, but not in his/her customary pool. According to Kuno (personal communication 2001) this means that the focus of negation (whether or not the focus is on the pool or the verb "swim") depends on whether the locative expression is readily interpretable as a member of a well-defined (contrastive) multi-member set, or whether it is a one-member set. Even the locative NP complement can become the focus of negation if uttered in contexts in which a choice exists between a pool and a lake, for example, as the place of swimming. That is, all that matters here is whether the locative NP complement can be interpreted contrastively or not. This naturally ties in with Kuno's claim regarding the scope of negation extended to a member of a multiple-choice set.

(47) a.

Fuji san	no	choojoo	wa	kugatsu	ni	shiroku-nakat-ta.
Mt. Fuji	NOM	summit	TOP	September	TEMP	white-NEG-PST

"The summit of Mt. Fuji was not white in September."

b.

Kyooshitsu	ni	terebi	ga	sanji	ni	nakat-ta.
classroom	LOC	TV set	NOM	3 o'clock	TEMP	exist-NEG

"The TV set was not in the classroom at three o'clock . . ."

The homogeneous predicates in sentences (47)a and (47)b are fundamentally not subject to the notion of "change." Therefore, when a punctual temporal NP complement is used in these sentences, it forces one to envision an instant that is particularly pin-pointed in distinction from the rest of the state. This interpretation causes one to identify the temporal NP complement as the focus of negation. When the temporal NP complement is not conceptually punctual, but durative, the focus of negation becomes less distinct and the sentence becomes less awkward. This is true for both the activity predicates used in sentence (46)b and the stative predicates used in sentences (47)a and (47)b, since both are concerned with the nonchange status of an event regardless of their temporal nonchange status or eternal nonchange status, respectively.

The behavior of various NP complements can be summarized as follows: punctual temporal NP complements in internal negation structures create awkwardness in the sentence and become the focus of negation regardless of the type of predicate, while durative temporal NP complements co-occurring with homogeneous predicates may create less awkwardness in the sentence and may not become the focus of negation; locative NP complements tend not to be the focus of negation when they co-occur with a verb with inherent direction, such as *iku* "to go," and *tsuku* "to arrive," but tend to serve as the focus of negation with an activity predicate; other nonobligatory NP complements, such as instrumental NP complements, usually become the focus of negation.

By investigating the focus of negation from a semantic viewpoint, it can be seen that negation in an internal negation structure, such as *kik-anakat-ta* "did not listen," essentially expresses an actual world situation, that is, a nonoccurrence of the event. Negative sentences that express actual world situations are objective, declarative statements that do not involve the speaker's conjecture about the truth or realization of the proposition. Simple negative statements are therefore outside of the domain of modality. Nevertheless, it has also been found that on many occasions NP complements accompanying a negative predicate cause a shift in the focus of negation, a shift from the negative predicate over to themselves. In such cases, the NP complement creates the implication that the actual event is the stated event's corresponding affirmative form. For example, the instrumental NP complement in *densha de ko-nakat-ta* "I did not come by train" brings the focus of negation on itself, which in turn conveys that the actual event is "did come." The sentence is well received if it is converted using an external negation structure. The negative form *ko-nakat-ta* "did not come" expresses a nonactual world situation that by all accounts would seem to be in the domain of modality.

Not surprisingly, however, the investigation of the scope of negation shows that this nonactual world situation exhibits a fundamental difference from the situations expressing modality that we have so far been investigating. Modality as examined through modals and conditional sentences was always found to be speaker-oriented. That is, modality is expressed on the basis of a speaker's beliefs and knowledge regarding the truth or realization of a proposition. The higher the degree of modality expressed, the more indeterminable the proposition in question becomes.

Nonactual world situations expressed by negative forms do indeed communicate the notion of indeterminability, but the key difference between them and actual world situations is that this indeterminability does not have its origin with the speaker. In these cases, where negative forms express nonactual world situations, modality is not based on a speaker's lack of firm evidence or confidence in the proposition's truth or realization. Here the truth or realization of a proposition is already known to the speaker. For example, when a speaker says *Sonoko wa ichigatsu ni umare-nakat-ta* "That child was not born January," the speaker already knows the actual world situation to be that the boy was born in February, for example. This is a case where a speaker implies an actual situation, and an interlocutor is left to infer whatever s/he will. As such, it is opposed to the situation where a speaker posits the truth or realization of a proposition. A speaker makes an implication when s/he knows the actual world situation. In this sense, the nonactual world situations expressed by negative forms are not possible/nonactual world situations from the speaker's viewpoint. Therefore, nonactual world situations expressed in internal negation structures are not within the domain of modality as defined in this body of work. This leads to the postulation that an external negation structure that clearly marks the scope of negation is not in the domain of modality. This also shows the fundamental difference in the use of the modals *wake* and *-no-da* from the suppositionals and evidentials. It should follow then that in external negation, implication on the part of the speaker causes an inference on the part of the interlocutor regarding the possibility of the proposition's truth or realization. This idea is taken up in the next section.

4.5.2 The Scope of Negation and Modality: External Negation

There are several expressions that can be considered to have an external negation structure in Japanese. They are found in Teramura's so-called "explanatory modals," (chapter 2) such as *wake dewa nai*, *-no-dewa-nai*, and *-to iu-no-dewa-nai*. These are equivalent to the English language expressions "it is not the case that . . ." or "it is not true that ~," which negate the part of a sentence following the sentential complementizer "that." The use of these modals clearly indicates that the semantics of these modals are different from other epistemic and deontic modals appearing in modal content.

In English internal negation, the constituents following after "not" all fall under the scope of negation, while the whole sentence is the scope of negation in external negation. For example, in the sentence "I do not eat sushi," "eat sushi" is the scope of negation. A parallel example in external negation is "it is not the case that I eat sushi," with "I eat sushi" as the scope of negation. In this external negation, there is an implication that the speaker does something with sushi other than eat it. In Japanese, although the scope of negation in an internal negation structure is not always clear-cut when NP complements are involved, in external negation the negative adjectival form *nai* affects the entire sentence in much the same manner as it does in English.[34] The following are examples:

(48) a. Kaw-anakat-ta
buy-NEG-PST
"(I) did not buy (it)."

b. Kat-ta-no dewa-nai.
buy-PST-COMP COP-NEG
"It is not the case that (I) bought (it)."

Sentence (48)a is simply a negative form of the predicate *kau* "to buy," and it expresses an actual event. If *kaw-anakat-ta* is not itself an actual event, such that the actual event was *morat-ta* "received (as a gift)" or *kari-ta* "borrowed," one should use the corresponding affirmative forms to express the actual event, or an external negation structure to imply the actual event, as illustrated in sentence (48)b, where the speaker is implying that the actual event was not that s/he bought it but borrowed it, for example. This indicates that the speaker utters this sentence from its basis as an actual event.

The following are cases when a sentence involves a temporal NP complement:

(49) a. Kurasu wa yoji ni owar-anakat-ta.
class TOP four o'clock TEMP end-NEG-PST
"The class did not end at four o'clock"

b. Kurasu wa yoji ni owat-ta-no dewa-nai.
class TOP four o'clock TEMP end-PST-COMP COP-NEG
"It is not that the class ended at four o'clock."

Sentence (49)a is an internal negation structure whose negative predicate expresses a nonactual event, and (49)b is the equivalent of (49)a in an external negation structure. These sentences can also be illustrated by the following diagrams:

34. In Japanese, the topic of a sentence is excluded from the scope of negation in both internal and external negation.

(49)a

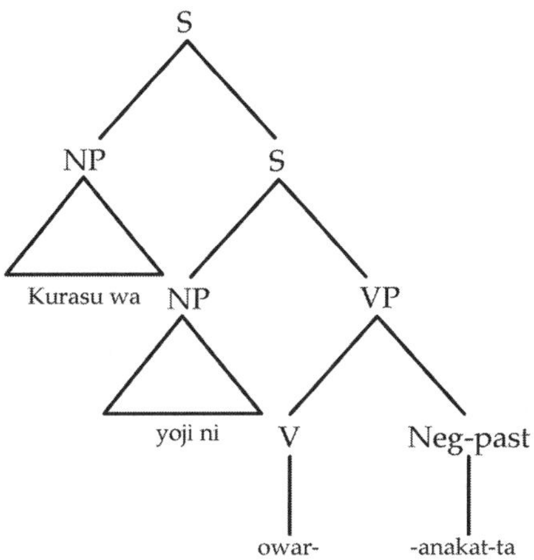

"The class did not end at 4 o'clock."

(49)b

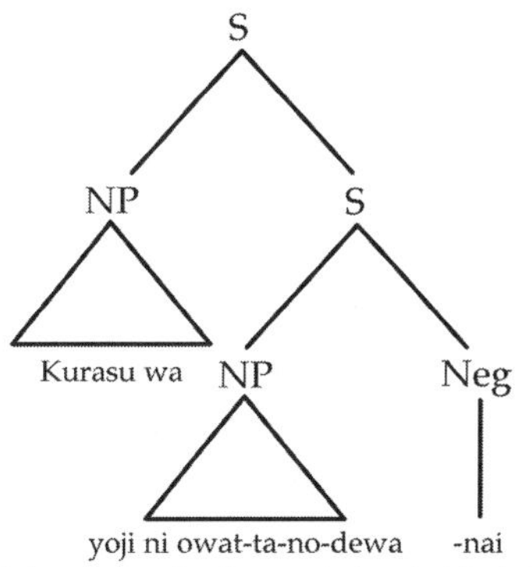

"It is not that the class ended at 4 o'clock."

Sentence (49)a implies that the class went on later than four o'clock. The external negation structure in (49)b clearly makes such an implication, although what happened in the actual world can only be contextually inferred (the same is true of internal negation). As is shown in the diagram, however, there is one noticeable structural difference between internal negation and external negation: while *nai* in an internal negation structure is dependently affixed to the predicate owaru "to end" and becomes an element of the negative predicate, *nai* in an external negation structure is an independent element in sentence structure. In external negation, then, either of the constituents *yoji ni* "at four o'clock" *or owat-ta* "ended" can be the focus of negation, whereas *yoji ni* tends to be the focus of negation in internal negation. An external negation structure creates many other possibilities for actual events, "The class ended later than four o'clock," "The class started (not ended) at four o'clock," or "The class started at five o'clock," to name a few. The more NP complements a sentence involves, the more possible interpretations the sentence has in its external negation structure. Here is an example that involves two NP complements:

(50) Kare wa uchi de sushi o taberu-no dewa nai.
he TOP home LOC sushi ACC eat-COMP COP-NEG
"It is not that he eats sushi at home."

In sentence (50), any of *uchi* "home," sushi, or *taberu* "to eat" can be the focus of negation. Multiple NP complements create the possibility of multiple interpretations. Thus, the speaker creates possible worlds for the interlocutor using external negation structures. This "speaker implies—interlocutor infers" relationship can be illustrated in structures involving the epistemic modal *kamoshire-nai* "may be." *Nai* in external negation structures possesses a function similar to that of epistemic modals that constitute modal content. The following diagram illustrates this:

(51)a

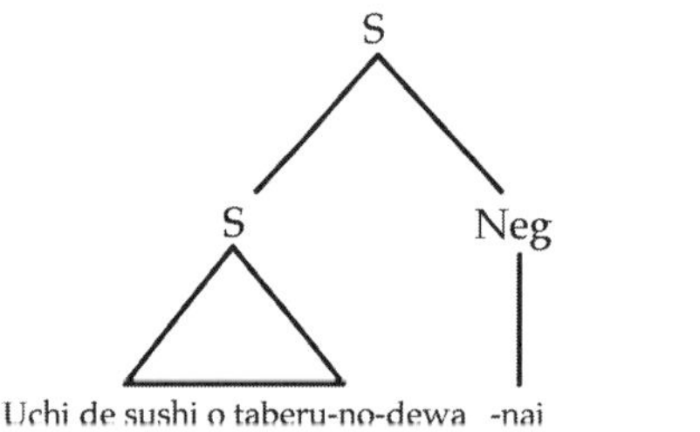

"It is not the case that (he) eats sushi at home."

(51)b

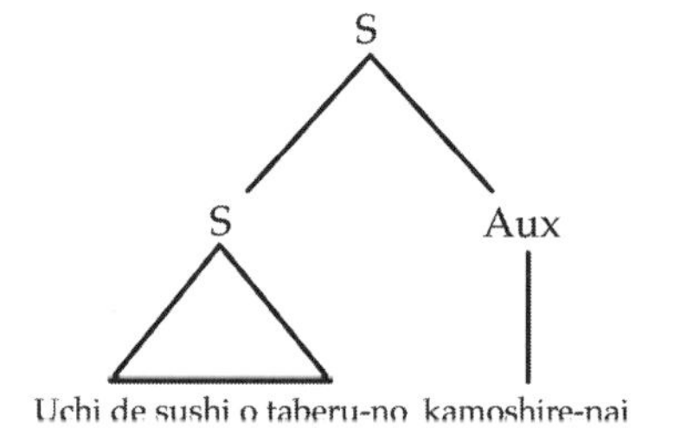

"It may be the case that (he) eats sushi at home."

As shown in the diagram, *nai* is located outside of the proposition *uchi de sushi o taberu* "eats sushi at home" as is the auxiliary *kamoshire-nai*. In this case, *nai* structurally constitutes modal content, which in turn suggests the existence of other possible situations. Diagrams (51)a and b reflect the speaker—interlocutor relationship in that when the speaker utters sentence (51)a, the interlocutor posits various possible world situations based on the speaker's utterance. Therefore, nonactual world situations expressed by negation are not possible/nonactual world situations where truth or the realization of the event is indeterminable from the speaker's view point. The use of an external negation structure is one communication strategy in which a speaker can cause an interlocutor to posit the possibilities of the truth or realization of other situations.

Many researchers (Tokieda 1950, Ishigami 1990, etc.) have noted that the morphology of *nai* allows it to vacillate between the expression of objective and subjective meanings. *Nai*, however, does not function in the same manner as modals (as illustrated above), and the nonactual worlds expressed by negation are suggested/implied worlds based on the speaker's actual world knowledge.

4.5.3 Summary: Negation and Modality

What has been discovered herein is that negative sentences are used to convey the nonoccurrence of an event as an actual event or, on the other hand, are used to imply the occurrence of some other event as an actual event. In either case, negative sentences were found to lie outside the domain of modality. Either negation structure, whether internal or external, invites one to envision a world other than that described. For instance, both the internal negation structure, *Sonoko wa yooka ni umare-nakat-ta*, "The child was not born on the eighth of the month,"[35] and the external negation structure, *Sono ko wa yooka ni umare-ta-no dewa-nai*, "It is not the case that the child was born on the eighth of the month," refer to the actual world situation that the child was born sometime other than the eighth of the month; therefore, the proposition, *yooka ni umare-ta*, "born on the eighth of the month," is within the domain of a nonactual world. This domain, however, belongs not to a world based on the speaker's conjecture, but a world that the speaker actually knows, inviting the interlocutor to infer the actual world situation.

Although it is not necessarily straightforward to detect what part of a sentence is the focus of negation in an internal negation structure (since it is primarily detectable through vocal impressions or context), NP complements in relation to the type of predicate play a significant role in shaping the determination. Heterogeneous predicates are fundamentally different in character from their homogeneous counterparts, and temporal NP complements quite often become the focus of negation. For heterogeneous predicates, the moment of change, which is usually indicated by a temporal NP complement, is different compared to any other moment in the time continuum. Once a change

35. Please note again that this sentence is, in fact, awkward. Sentences that employ the internal negation structure and whose focus is other than the predicate are usually unstable.

occurs, all moments in the resultant state are purely homogeneous. Therefore, when a temporal NP complement co-occurs with a negative heterogeneous predicate, the nonoccurrence of the event expressed by the negative heterogeneous predicate at a particular moment automatically suggests the occurrence of the event at some other time. It is difficult if not impossible to view the nonoccurrence of an event at a particular moment. This is evident in the interpretation of sentence (45)b, *Sono ko wa yooka ni umare-nakat-ta*, "that child was not born on the eighth of the month," in that one can not envision the nonoccurrence of the child's birth at a given moment; rather, the interlocutor immediately infers the occurrence of the event at some time other than the eighth of the month. For this reason, the external negation structure allows the interlocutor to envision the actual situation more directly.

Similarly, homogeneous predicates, such as activities and states, are semantically not good counterparts to temporal NP complements. This is attributed to the fact that both activities and states involve the notion of duration. Even though a given temporal moment is among those that constitute the entire duration, pin-pointing a particular moment forces one to envision the rest of the consistent homogeneous state occurring at some other time; therefore, a durative NP complement is more appropriate than a temporal NP complement in order for the interlocutor to more directly perceive the nonoccurrence of the event.[36]

Thus for internal negation structures, temporal NP complements generally become the focus of negation for both types of predicates, and such complements in combination with heterogeneous predicates in an internal negation structure create awkwardness. Instrumental NP complements can also be the focus of negation, since they are not semantically rooted in predicates. Locative NP complements, on the other hand, tend not to be the focus of negation, since they are semantically rooted, semi-obligatory NP complements associated with motion, activity, or existence verbs. The external negation structure provides more possible interpretations as to which part of the sentence is the scope of negation.

36. The duration of a temporal NP complement is judged not by the actual time required for an event to be realized, but by the environment of the sentence in which the complement is used. For example, the year 1989 is not punctual when compared to today, but it can be considered as punctual when a person's entire life is considered.

The behavior of these NP complements, in fact, coincides somewhat with Kuno's notion of "multiple choice." As pointed out by Katoo (1985) and others, the term "multiple choice" does not give a clear picture of what becomes the object of choice. This idea cannot be resolved from a syntactic viewpoint, but is comprehensible from a pragmatic viewpoint. When writing a letter, for example, it is commonly understood that a person needs a pen, a pencil, or some other writing utensil. When the mention of such devices co-occurs with a negative predicate, such as in "I did not write the letter using a pen," one does not have a difficult time calling other devices to mind. For example, we may assume that s/he wrote the letter using a pencil instead of a pen. Information is derived based on common knowledge in a particular culture (e.g., the device normally used to write a letter may differ from culture to culture). If we do not possess knowledge of the device presuppositionally, a negative statement that tries to call that particular device to mind will simply fail. For example, the statement "I did not write the letter using a red pen" probably conveys to the interlocutor that the speaker used a different color pen. The sentence does not suggest for example, that the speaker used "orange juice" to write the letter. Since orange juice is not a commonly accepted component of letter writing, the negative statement does not succeed in conveying that implication. A straightforward affirmative statement such as "I used orange juice to write the letter" is required to effectively communicate such.

Kuno's multiple choice notion is applicable to such instrumental NP complements, since they are relatively restricted in terms of conceivable alternatives. For example, one can "go" by walking, taking a car, bus, bicycle, train, airplane, and so forth. When someone says, "I did not come by bus," the purpose of the statement is usually to imply how s/he did arrive or to convey that, contrary to some expectation, s/he did not take the bus on this occasion. Such corresponding affirmatives, however, rely not on universal language phenomena, but rather on knowledge common to a particular culture. Kuno's notion of "multiple choice" then becomes a question of pragmatic domain. It is difficult to precisely define the notion of multiple choice, since it is not merely a question of syntactic features, but requires one to construct a corresponding affirmative based on "common knowledge."

The idea of presuppositional knowledge leads to Givon's claim (1975) that negative sentences generally occur where the corresponding affirmative has been mentioned or contemplated, or when the speaker believes that the interlocutor tends toward the affirmative. Examining affirmative predicates appears to be crucial to investigating negative predicates, that is, negative statements cannot be effective without reference to corresponding affirmative statements. A negative predicate accompanied by an instrumental NP complement is employed when a speaker uses some "understood" fact(s) to obliquely convey information more obviously expressed by an affirmative form of the statement. If one does not possess the information expressed by the affirmative form, the negative form cannot play its role.

The impracticality of negation expressing modality is supported by Givon's postulation that presuppositional knowledge of the speaker affects the use of a negative sentence. In nonactual world situations expressed by negative sentences, a speaker is presenting an actual world situation in a suggestive, periphrastic manner in which the interlocutor makes a judgment concerning the truth or realization of the expressed proposition. Thus, an external negation structure is a pragmatic device that can be used to make a purposely ambiguous remark rather than offering a clear picture of the actual situation. The interlocutor is left to make a judgment based on the implication the speaker has given.

Negation is thus revealed as nonmodal in the manner that modality is described herein. Even though the syntactic behavior of *nai* in the external negation structure is identical to that of suppositionals and evidentials, the semantic implication of *nai* is clearly unequal to that of such epistemic modals; rather, it is the same as Teramura's explanatory modals, such as *-no-da* and *wake-da.* The nonactual worlds expressed by explanatory modals as well as by negation are suggested/implied worlds based on the speaker's actual world knowledge, and, therefore, the indeterminability brought about from the sentence does not have its origin with the speaker. The investigation of the behavior of *nai* also demonstrates the difference between epistemic and explanatory modals.

4.6 SUMMARY OF MODALITY AND ASPECTUAL CATEGORIES

From our examination of aspectual categories, it becomes clear that the notion of "change" is central to the notion of aspect, involving two different situations—one before the change occurs and the other after the change is brought about on the same time continuum. These two situations become distinct through the use of the aspectual form *te-iru.* When *te-iru* is used for homogeneous predicates, it brings two different aspects into these two different situations—progressive and resultative (perfect). When *te-iru* is used for heterogeneous predicates, only the latter meaning is applicable. For both types of predicates, resultative (perfect) aspect shares the same features as any predicate compatible with *te-iru.* A resultative state refers to a particular period, one located between the time that change last occurred and the time that another change started or may start. This resultative state is a pure static state, one which does not require outside energy to sustain it; it is a state for which volitional interference on the part of the speaker is out of the question. Therefore, a resultative *te-iru* presents pure stativity that is represented by adjectives, nominals, and stative verbs.

The term "stative" is employed to refer to a static state in which no energy input is required to maintain the state, essentially equivalent to the definition proposed by Comrie (1976). In contrast are dynamic events that exhibit clear points of initiation and/or termination on a time continuum. Stativity can be recognized as a concept, but cannot be experienced as a dynamic event. For example, "beautiful" is merely a concept, and one cannot experience "beautiful" in and of itself, although one can experience "seeing a beautiful painting." According to Soga (1983), stativity refers to a nonself-controllable state that exists beyond a speaker's volitional control and has various selectional restrictions in its syntactic and semantic behavior. Jacobsen (1990b) also discussed the multifaceted character of stativity from the viewpoint of five categories of meaning that interact with stativity, claiming that states denote the absence of any element of change in their meaning and represent single homogeneous situations within every subinterval holding exactly the same quality. According to Dowty's premise, the homogeneous category that denotes an absence of change views states and activities as within the same group. Both states and activities share a common feature in that they reflect an absence of change, yet they are dissimilar in that activities are self-controllable and require constant energy input in order to maintain the activities.

Stativity is thus concerned with worlds over which a speaker has no volitional control, and no energy input by a sentient being can bring change to the situation. The negative suffix *nai* was also discussed in terms of aspect. Since *nai* is morphologically an adjective, the assumption is that *nai* has the effect of converting nonstative predicates to stative predicates.

The examination of aspectual categories revealed that both aspectual categories share common functions. One such commonality is the expression of a situation where energy is applied, as was seen in the appropriateness of *nai* co-occurring with *te-iru. Nai* was shown to behave in ways very similar to the original affirmative form of the verbs of volitional activity. This demonstrates that *nai* as a suffix is not necessarily sufficient to convert the activity event to a static event.

Another common function is the representation of a pure static state where no energy input is required. This state is purely homogeneous, with every interval of the subinterval state sharing exactly the same property. For *te-iru,* this property comes from its ability to express the perfect state resulting from an influx of energy that converted some previous state into the resultative state. For *nai,* pure homogeneity is seen in the fact that *nai* functions as an adjective (as seen in *okane ga nai* "money does not exist/I do not have money") or an adjectival suffix (as seen in *ik-anai* "I will not go") in a simple sentence. This bifunctionality of *nai* communicates Tokieda's postulation (1955) that *nai* vacillates between objectified notions and subjectified notions. The former are considered to be within the domain of propositional content, whereas the latter are considered to be within the domain of modality in which *nai* can form part of the external negation structure.

Te-iru and *nai* then are included under the concept of "stativity" in that they can depict either an eternal phenomenon containing no clear points of inception/terminus or a resultative state lying between periods of energy influx. Stativity, expressed by these grammatical forms, represents the conceptual world where there is no interference from volitional control; it creates a spatiotemporal distance between the speaker and the state itself. The concept that stativity generates at root is relevant to the basic notion of modality in that a higher degree of modality is expressed in the form of possible/nonactual worlds where a speaker can merely hypothesize the realization or truth of the proposition, that is, worlds beyond a speaker's volitional involvement and manipulation.

Chapter 5

Modality and Conditionals

To explore a possibility is simply to consider an alternative way the world might have been or might be, that is, a nonactual situation. Such is the case with conditional sentences, since conditionals generally invite one to imagine that certain hypotheses hold, and invite consideration of the consequences of these hypotheses. They do not, however, require the speaker to commit to the belief that a given proposition is true. They have neither truth value nor factivity, but express possible worlds in which one cannot determine the truth of the antecedent nor the truth of the consequent. In this sense, conditional sentences can be said semantically to express modality, but they are syntactically in the domain of propositional content as they can accompany a modal.[1]

This chapter concerns careful scrutiny of the degree of hypotheticality expressed by three Japanese conditional forms—*to, ba,* and *tara*—which together constitute roughly the same meaning as "when,"

1. Following are some examples that support this premise:

(a)

Konna	koto	iu-to	shitsureini-naru	kamoshiremasen	ga . . .
this kind	thing	say-COND	rude-become	AUX	but

"It might be rude to say things like this, but . . . ."

(b)

Yuki	ga	fur-eba	kare	wa	ko-nai	ni chigainai.
snow	NOM	fall-COND	he	TOP	come-NEG	AUX

"I'm sure that he will not come if it snows."

"whenever," and "if" in English.[2] "When" is used to imply certainty of the occurrence of the event expressed in the antecedent clause; "whenever" indicates that the event in question occurs/occurred repeatedly, thus resulting in predictability of the event's realization from the speaker's viewpoint. These two meanings are concerned with the notion of epistemic necessity in that the speaker is confident in the truth of the events described by the conditional sentence. "If," on the other hand, serves to signal the speaker's supposition of the antecedent event, and the truth of the described event is unknown to the speaker; therefore, "if" is concerned with the notion of epistemic possibility. Whether a Japanese conditional sentence expresses "when," "whenever," or "if" depends on the nature of the consequent clause, and on pragmatic judgments made by the speaker. In this sense, as used by Hinds and Tawa (1975), the term "conditional" refers to the complete sentence—including both the antecedent and the consequent—that contains any of the conditional forms. Conditional sentences thus involve a causal relation between the antecedent and the consequent. The following sentences exemplify this relation:

(1) a. Honda san ga kuru-to . . . .
Mr. Honda NOM come-COND
"When/whenever/if Mr. Honda comes . . . ."

b. Honda san ga ki-tara . . . .
Mr. Honda NOM come-COND
"When/if Mr. Honda comes . . . ."

c. Honda san ga kure-ba . . . .
Mr. Honda NOM come-COND
"If/whenever/when Mr. Honda comes . . . ."

All three examples are antecedents meaning "when/if/whenever Mr. Honda comes." The meaning of each antecedent, however, can-

2. *Nara* may also be counted among Japanese conditionals, but contrary to the function of the other three conditionals, *nara* does not possess a bipolar function and expresses only an "if" meaning. Also, a hypothetical interpretation of *nara* is often identified based on previous context. *Nara* is therefore outside the subject of investigation in this chapter. For detailed discussion, refer to Johnson (1997).

not be determined without a consequent. For example, sentence (1)a can be interpreted as "when" or "whenever" when the consequent is, for example, *itsumo naga-banashi o suru* "I always chat for a long time." Alternatively, the sentence can be interpreted as "if" when the consequent is, for example, *shinsoo ga wakaru kamoshire-nai*, "we may find out the truth." Also, if the consequent of (1)c refers to a situation in the past, such as *naga-banashi o shi-ta mono-da*, "I used to chat for a long time," then the English equivalent of *ba* is "whenever." Thus, the Japanese conditional forms have the capacity to express either a nonhypothetical or a hypothetical situation depending on the consequent. Such language use is in part due to the fact that, for both temporal "when" and conditional "if," the most frequent relation between antecedent and consequent is causal in nature.

According to Reilly (1986, p. 310), numerous languages exhibit only a single morpheme for some of the functions served by "if" and "when," such as the German *wenn* in present and future tenses. She demonstrates that a "when" temporal and an "if" conditional clause are roughly synonymous, even in English, as in "If/When Mary takes aspirin, she gets a rash" and "If/When it rained last year, the west bank of the Mississippi flooded." Reilly examined such usage employed by children, concluding that children generalize the semantic functions to other linguistic forms and complementarily use alternative forms for similar semantic functions in their acquisition of conditionals (p. 328). Employing the view of conditional sentences proposed by Schachter (1971), in which there are two types of conditionals (simple and imaginative), Reilly further examined cases in which "when" and "if" are mutually appropriate. Simple conditionals refer to events in the real world and include present, past generic, and predictive conditionals, while imaginative conditionals reflect some notion of irrealis, and they include both the hypothetical and the counterfactual. Among these conditional sentences, predictive conditionals also allow the sentence to be restated using "when." Accordingly, children distinguish "when" and "if" based on their expectation of the antecedent's occurrence. This can certainly be applied to adult language use as well. Therefore, the motivation for using particular conditional forms is born from the information possessed by a speaker, and which interpretation a sentence receives—nonhypothetical or hypothetical—depends on what a speaker has managed to convey. Such a speaker-oriented notion in the use of conditionals is a common feature of many languages. Akatsuka (1986) similarly concluded that "conditionals do not belong to the static domain of mathematical logic" (p. 333), but are "discourse-bound because the fundamental question, 'What is a conditional in natural languages?' itself is unanswerable without postulating a specific speaker and that speaker's attitude towards the state of affairs expressed by the antecedent" (p. 349).

Akatsuka (1983), however, argues that forms that express nonhypotheticality should be excluded from the conditional category, since it has been assumed that a true conditional must at least be hypothetical in meaning. She claimed that conditionals can be arranged on a scale in the domain of irrealis (nonactual worlds); therefore, when conditional forms receive a nonhypothetical interpretation they cannot be regarded as conditionals. When the consequent expresses a past event, the sentence is in the domain of actual worlds, but when the consequent expresses a nonpast event, this is not the case. This means that Japanese conditionals sometimes qualify as conditionals and sometimes do not, depending on the causal relation and upon the environment in which the speaker exists. Please note the following examples:

(2)

Kabu	ga	kyuubooraku-suru-to
stock	NOM	sudden fall-COND

konohen	no	shinkoo bijinesu	wa	mikka	mot-anai.
this vicinity	NOM	new business	TOP	three days	last-NEG

"When/if the stock falls suddenly (in the Japanese economy), new businesses in this vicinity will not last even three days."

(3)

Kekka	ga	de-tara	oshiete-kudasai.
result	NOM	come out-COND	tell-please

"If/when the result comes out, please let me know."

As seen in example sentences (1)a, b, and c and in sentences (2) and (3) as well, whether a sentence is intended to convey "when" or "if" is determined by the speaker's perception and knowledge. Therefore, the hypotheticality or nonhypotheticality of these sentences cannot necessarily be determined by the interlocutor. For example, if sentence (2) is spoken under the assumption that the deterioration of the Japanese economy is considered highly probable among people in the community, the sentence gains a "when" interpretation. The interpretation of sentence (3) is indeterminable unless further context is provided. It depends on the speaker's perception of the person from whom the speaker requested the answer. Under Akatsuka's premise, therefore, one needs to examine the environment of every single sentence in order to determine which meaning the respective Japanese conditionals express.

Alternatively, Jacobsen (1992a) claimed that the fundamental feature of conditionals should be viewed in terms of "contingency" rather than "nonhypothetical modality." He regarded conditional sentences as "expressing a relationship between two events whereby one event is contingent on the prior realization of another," (p. 6) and treated the question of whether the antecedent event is actually realized or not as an independent feature. Taking advantage of this approach and viewing the conditional forms in terms of "contingency," one can examine both hypothetical and nonhypothetical meanings in the domain of conditionals. Please note the following examples that include the conditional forms, *to, ba, and tara:*

(4) a. Nishi no sora ga akai-to tsugi no hi wa hareru.
west GEN sky NOM red-COND next GEN day TOP sunny
"When/if the sky in the west is red, the next day will be sunny."

b. Nishi no sora ga akakat-tara tsugi no hi wa hareru.
west GEN sky NOM red-COND next GEN day TOP sunny
"If/when the sky in the west is red, the next day will be sunny."

c. Nishi no sora ga akake-reba tsugi no hi wa hareru.
west GEN sky NOM red-COND next GEN day TOP sunny
"If the sky in the west is red, the next day will be sunny."

All three conditional forms are appropriate in the example sentences. The degree of hypotheticality each sentence expresses, however, is not exactly the same. Sentence (4)a gives us the impression that the idea "when the sky in the west is red the next day will be sunny" is a generally accepted fact. A nonhypothetical interpretation is more appropriate here than a hypothetical interpretation. Sentence (4)b, however, may be used in a situation where the speaker is referring to a particular day, such as in this case: "If the sky in the west is red today, tomorrow will be sunny." In this situation, the resulting interpretation will be hypothetical slightly more often than it will be nonhypothetical. Sentences (4)c invites an "if and only if" interpretation, rather than a nonhypothetical interpretation.

People with knowledge of a language exploit language functions without necessarily paying careful attention to issues such as grammatical, and phonological, sociocultural accuracy, in conveying their thoughts. Choice of vocabulary and grammatical patterns are employed according to how a speaker views the situation, and this can only be observed externally. For example, it may be the case that a speaker uses *to* when she knows or views the situation to be a generic, habitual, or objective fact, and she uses *ba* when she perceives the situation as hypothetical. The functions of the language are already internally acquired by a native speaker, but they are not yet externally analyzed. The analysis becomes problematic when two distinct meanings expressed in a language are expressed by a single morpheme in a different language. This is exactly what Japanese conditionals are about. Speakers employ conditional forms to convey their thoughts accurately, yet when interpreting them in a different language, the interpretation sometimes creates a meaning that the speaker did not intend.

In investigating conditional forms, most researchers state that the conditionals *to* and *tara* are used to express both hypothetical and nonhypothetical meanings since they can be used to express past events. *Ba,* on the other hand, is commonly understood to express only hypothetical situations due to the defining characteristic that it cannot be used to express past events. As Dancygier (1998) points out, "One solution that fits all kinds of approaches is not common among linguists, because a linguistic analysis cannot fail to notice the significant differences between types of conditionals" (p. 4). The next section carefully examines the degree to which these characteristics affect the interpretation of hypotheticality, thereby revealing the differences between the Japanese conditionals.

5.1 HYPOTHETICALITY AND THE *TO* CONDITIONAL

In English, the use of "if," "when," or "whenever" depends on the speaker's knowledge about the realization/truth of the antecedent, and that knowledge is unambiguously presented to the interlocutor right at the beginning of the sentence. Such is not the case in Japanese, in which a degree of ambiguity is present, since one must discern a speaker's internal, mental dynamics as reflected by the grammatical devices employed by the speaker and the manner and environment in which the sentence is uttered.

Alfonso (1966) and Kuno (1973) are representative of the many researchers who have concluded that *to* conditional sentences are used to express generic, habitual, objective facts and a temporal connection between the antecedent and the consequent. The justification is that when the content of a conditional sentence expresses a universal truth, *to* is the most appropriate choice from among the three conditional forms. Taking this for granted, some researchers claim that *to* sentences express only generic, nonhypothetical situations.

5.1.1 *To* Sentences in Nonpast Context

5.1.1.1 Nonhypothetical To

Generalized facts are usually acquired through observation of some common, iterated result. For example, people may come to know by direct observation of periodic seasonal changes that snow starts falling sometime in October in Minnesota. A repetitive experience like this comes to be eventually registered in the mind/body as knowledge, skill, or a practice derived from direct observation of, or participation in, such events. Once these experiences are registered, the consequences of like situations are easy to predict. A *to* sentence is used in a situation where a speaker finds the consequences of an event predictable, that is, a situation that does not require one to hypothesize. The following example sentences represent the use of *to* sentences in nonpast context:

(5) Yoru ryoku-cha o nomu-to nakanaka nemu-re-nai.
night green tea ACC drink-COND easily sleep-POT-NEG.
"When/if I drink green tea at night, I cannot fall asleep easily."

(6) Mizu o koor-aseru-to taiseki ga fueru.
water ACC freeze-CAUS-COND volume GEN increase
"When/if one freezes water, the volume increases."

(7) Kono shiken ga owaru-to yasumi ni hairu.
this exam NOM finish-COND break LOC enter
"When this exam is over, we will have a break."

(8) Juugatsu ni-naru-to Minesota de wa yuki ga furu.
October-become-COND Minnesota LOC TOP snow NOM fall
"When October comes, snow falls in Minnesota."

Sentence (5) describes an individual's practice and gives the impression that being kept awake by green tea is a habitual event, a fact that the speaker has learned from iterated experience.[3] Therefore, the whole sentence can be viewed as expressing nonhypotheticality—in this case, habituality. The same observation can be applied to sentence (6). The difference is simply whether or not the predictable event is valid only to a single individual or to a large number of individuals. Whereas the notion expressed in sentence (6) transcends cultural or individual differences—that is, it is a general fact—of course, the statement in sentence (5) may apply only to a given individual,

3. Comrie (1976) refers to the possible confusion created by the use of the terms "habituality" and "iterativity." He defined "habituality" as describing a situation that is characteristic of an extended period of time, so extended in fact that the situation referred to is viewed not as an incidental property of the moment, but as a characteristic feature of a whole period. Hence, there is no need to involve or exclude iterativity. For example, the sentence "He sneezed five times" is perceived as an iterated activity, but this type of repetition cannot be considered habitual. Also, the notion of habituality can be established without iteration, as seen in, for example, "The government office building of Tokyo used to be at Marunouchi." Thus, iteration is not necessarily habitual, and habit is not necessarily born of iteration.

and would not necessarily apply to a larger group. In fact, it may be valid only for the speaker. Still, the predictability of the event to the speaker allows him to use the to construction, and each of these sentences expresses a nonhypothetical situation.

A question arises, however, as to why the English equivalent can be "if" as indicated in the translation. Both *to* antecedents include an activity that can be volitionally controlled by the speaker. In both cases, the speaker can hypothesize the realization or the truth of the antecedent. For example, sentence (5) can be perceived as "the event of my drinking green tea may not be realized at the time of speech, but supposing that it is realized, then I know that such action results in a situation where I cannot sleep easily." In order to receive either a "when" or "if" interpretation in English, the event in the antecedent has to be one that can be approached from either the realization of the event in the future being known to the speaker or the realization of the event being yet unknown to the speaker.

Contrary to the above observation, sentences (7) and (8) are consistent only with a "when" interpretation, since the unpredictability of the event's realization is not compatible with the described situation. The antecedent of sentence (7) depicts a specific occasion in which a particular exam will be finished, and the speaker takes cognizance of the vacation's consequent forthcoming. Neither a "whenever" interpretation (that is incompatible with specificity) nor an "if" interpretation (that is incompatible with the speaker's knowledge) is applicable, but a "when" interpretation (that expresses the speaker's knowledge of a particular event) is applicable. Sentence (8) describes a general event and cannot be perceived in the same fashion as sentences (5) or (6), since "to become October" cannot be in the domain of a hypothetical world. Sentence (8) is distinct from sentence (7), however, in that "to become October" is not a specific event, but one accepted as a universal truth. As far as the nonfictional, actual world in which we live is concerned, it comes every year as a generalized fact, and in both languages, only the nonhypothetical interpretation ("when," considered as an event out of every coming year) is appropriate. Situations described by *to* sentences are thus in the domain of nonhypothetical worlds—that is, a speaker finds the consequence predictable.

Although realization of future events is not objectively perceptible, the predictability of events creates a nonhypothetical "when" interpretation. Given such interpretation, the temporal connection between the antecedent and consequent is clearly identified: "when/whenever *p*, then *q*." Therefore, the simultaneous occurrence of two events cannot be described by the *to* conditional, as exemplified in the following sentences:

(9) *	Kuruma	o	untensuru-to	megane	o	kakeru.	
	car	ACC	drive-COND	glasses	ACC	wear.	
	"When I drive, I wear glasses."						
(10) *	Yukimichi	o	aruku-to	kono	buutsu	o	haku.
	snow street	ACC	walk-COND	this	boots	ACC	wear
	"When walking on the snowy street, I wear this pair of boots."						

The inappropriateness of sentences (9) and (10) is attributed to the fact that the two events in the sentence are occurring concurrently.[4] The speakers wear a pair of glasses or boots at the same time as when s/he drives the car or walks in the snow, respectively. However, the interpretation of "when *p*, then *q*" (i.e., a causal relationship) cannot be identified in the events in sentences (9) and (10) even if these events are habitual, generalized facts.

According to the above view of general events, however, if an experience expressed in a *to* sentence is merely a one-time event[5] or an unfamiliar event for the speaker, the experience may not be internalized mentally as a fact.[6] The experience is not sufficiently stable to convince us that the given situation will always result in the same consequence. Speaker volition, for example, does not convey a guaran-

4. In this case, a *toki* clause may be used, since *toki* can refer to a point in time when an activity starts or a change occurs. *Kuruma o unten suru toki megane o kakeru* means "I put on glasses when I start driving the car," with the subordinate clause indicating the point in time at which the speaker dons his/her glasses. On the other hand, to refer to the duration of time in which the speaker is wearing glasses (e.g., throughout the period in which he/she is driving), *aida* "while" may be used.
5. In English, "when" can be used to refer to one-time, unique events or habitual events such as "when I get home, I will call you" and "when graduation day comes in spring, people walk around naked in Ann Arbor," respectively. Both notions are within the domain of nonhypothetical worlds, and which meaning the sentence expresses is based on the pragmatic judgment of the content of the antecedent.
6. Scientific and mathematical facts may be internalized without iterated experience. We do not hypothesize regarding such facts, since they have been conveyed from one person to another since the time they were found to be facts.

tee of a particular realization of events, since human volition lacks accountability, generality, and universality; therefore, the outcome of volitional events[7] is not predictable, and sentences that express human volition in the consequent are grammatically incompatible with *to* constructions. The following examples demonstrate the validity of this statement:

(11) *	Ashita	ame	ga	furu-to	doo-suru.
	tomorrow	rain	NOM	fall-COND	how-do

"What are we going to do if it rains tomorrow?"

(Koide, Komatsu, and Saita 1981)

(12) *	Ashita	hareru-to	baabekyuu	o	shi-yoo.
	tomorrow	sunny-COND	barbecue	ACC	do-VOL

"If it is sunny tomorrow, we shall barbecue."

Although the English translations of sentences (11) and (12) are grammatically acceptable, such is not the case for the original Japanese. Both of the above sentences express the volition of the speaker: the consequent *doo-suru* "what are we going to do" in sentence (11) and "we shall barbecue" in sentence (12) express the speaker's volitional plans depending upon the next day's weather. In sentences (11) and (12), the relationship of the antecedent and the consequent is based neither on observable facts nor on iterated experience. Human volition cannot support the establishment of a generally valid fact, and there is no guarantee of an actualizable, temporal connection between the antecedent and the consequent. The unacceptability of the speaker's volitional involvement in the consequent is therefore attributable to the conflict created by the fundamental function of *to*. Careful incorporation of such a notion in translation of these sentences results in English equivalents that are also considered awkward—for example, "What are we going to PREDICT TO DO if it rains tomorrow?" for sentence (11), and "If it is sunny tomorrow, I PREDICT THAT we shall barbecue" for sentence (12). It is clear that *to* sentences

7. Volitional events are represented by expressions that indicate a speaker's will, desire, intention, request, command, and so forth.

are used to convey the speaker's knowledge about nonhypothetical situations gained through iterated observation or acquired as scientific facts. In this sense, *to* is not in the domain of modality.[8]

This observation does not imply, however, that *to* sentences show no involvement with hypothetical worlds. Although the fundamental function is the representation of nonhypothetical worlds, there are certainly future events that are sufficiently stochastic (i.e., unpredictable) that this feature of *to* is overturned, and the sentence expresses hypotheticality. After all, as taken up in the following section, future events are subjects of unknown worlds that overall remain to be seen.

5.1.1.2 Hypothetical To

The fewer chances an event or phenomenon has to be internalized as a general fact, the more hypothetical the consequences of that event become. Conversely, the more chances, the less hypothetical the consequences. For example, imagine a time line with a dot on it representing a particular experienced event. Each time a similar event is experienced, another dot is added to the line to represent that experience. The more dots on the line, the less hypothetical the consequences of the event. Since predictability increases with the frequency and/or the total number of the experiences, when the dots become so frequent or numerous as to appear as a line themselves, the dots represent a habitual event. Also, when the view of this same line is shared by all people, the dots represent a universal truth, that is, a fact. Therefore, once an event becomes a fact, that event is always represented as a fact at any moment of a time interval, that is, the fact represents pure homogeneity. Universal truths are thus perceived as not being limitable to one specific point in time, but rather have to be held as truth at any moment. This is why eternal truths are expressed in nonpast tense forms, which themselves are used to indicate the semantic notion of future or habitual events in a simple sentence. This explains why nonpast tense forms are used in *to* conditional sentences. The expression of nonhypotheticality is thus based on experienced, internalized, and generalized events, while the expression of hypotheticality is based on unexperienced, uninternalized, and indeterminable events, all three of which are in the domain of possible/nonactual worlds.

8. Inaba (1991) also examined the functions of *to* in terms of the broad sense of modality, and demonstrated that *to* does not occur with modality expressions (volitional expressions herein) and stated that *to* is in the domain of [-Modality].

Examining events in the domain of possible/nonactual worlds, therefore, should offer an indication of whether *to* sentences may express hypotheticality at all. Some researchers, such as Kawaguchi (1984) and Koide, Komatsu, and Saita (1981), have asserted that *to* does not function as expressing a hypothetical meaning, while some others, such as Eda (1991), demonstrated that *to* can be used to express a hypothetical meaning as high as 13%.

The notion that is related to unexperienced, uninternalized, and indeterminable events is best related to the speaker's volitional uncontrollability of an event. Stative predicates, nonvolitional unaccusative verbs, predicates that describe another individual's activity, and natural phenomena are all (at least somewhat) outside the domain of speaker control; therefore, the speaker cannot predict the realization of the event in the antecedent.[9] The following are some relevant examples:

(13) Ashita hareru-to ii naa.
tomorrow sunny-COND good PART
"It will be good if it is sunny tomorrow (I hope it will be sunny tomorrow)."

(14) Anata ga ik-anai-to minna ga komaru.
you NOM go-NEG-COND everyone NOM trouble
"If you do not go, everyone will be troubled."

(15) Kane o das-anai-to shinjitsu o barasu zo.
money ACC give-NEG-COND truth ACC expose PART
"If you do not give me some money, I will expose the truth."

(16) Uchi no musuko wa, moo sukoshi hansamu-da-to
family GEN son TOP more little handsome-COP-COND

moteru-n-da kedo nee . . .
popular-COMP-COP but PART
"As for my son, if he were a little more handsome, he would be popular among girls."

9. The above statement is true unless, again, the event pertains to a scientific phenomenon that can be completely internalized and learned from a single experience.

Sentences (13) to (16) cannot be identified with either a "whenever" or a "when" meaning even though one can envisage sentences (13), (14), and (15) as one-time events; they all express hypotheticality. This is conceivably attributed to the fact that the properties of these events do not grant a direct, materialized experience upon which the speaker could generalize, since human volition cannot manipulate the properties of the events. Therefore, these events tend to overturn the fundamental function of "*to*," that is to capsize the nonhypothetical meaning to a hypothetical meaning. Stative predicates, nonvolitional unaccusative verbs, predicates that describe another individual's activity, and natural phenomena thus play a significant role in the expression of hypotheticality by *to*. They all create a spatiotemporal distance between the situation and the speaker that are in the domain of possible/nonactual worlds, that is, they help *to* to express modality.

In nonpast context, *to* sentences are used to express three meanings—habitual/general (whenever), one-time (when), or hypothetical (if)—depending on the situation and the type of predicates used in the *to* construction. In fact, the ability of *to* to express hypotheticality in nonpast context should be considered justifiable, since the future is not necessarily predictable even though predictable cases can often be observed through repeatedly experiencable events and scientifically proven facts in the real world. In past context, however, interpretations of *to* sentences may not be exactly the same as those in nonpast context, since past events are already realized and do not allow one to hypothesize upon the outcome.

5.1.2 *To* Sentences in Past Context

5.1.2.1 Nonhypothetical To

Predictable, nonhypothetical, objective facts expressed in *to* sentences are concerned with actual worlds since they refer to observed events. This is related to the fact that *to* sentences can also be used to express past events. Facts described as past events are objectively perceptible, since they no longer belong to a world under human control. The following examples illustrate the use of *to* in past context:

(17) Juuichigatsu ni-naru-to sassoku yuki ga furi-hajime-ta.

November become-COND right away snow NOM fall-begin-PST

"Snow began to fall soon after November came."

(18)

Tonneru	o	nukeru-to	soko	wa	yukiguni dat-ta.
tunnel	ACC	go through-COND	there	TOP	snow country-COP-PST

"The train came out of the long tunnel into the snow country."

(Yasunari Kawabata's "Yuki-Guni" [1948]
translated by Edward Seidensticker [1981])

(19)

Kazuo	to	me	ga	au-to
Kazuo	COM	eye	NOM	meet-COND

Sonoko	wa	kimatte	soppo	o	mui-ta.
Sonoko	TOP	without fail	different direction	ACC	look-PST

"When Sonoko's eyes met Kazuo's, she always turned away."

Even though sentences (17) and (18) express particular events, and sentence (19) expresses a past habitual event, all three sentences describe facts in past situations and express temporal connections between the antecedents and the consequents. These *to* sentences simply express sequential events, that is, *p*, then *q*.[10] Sentences (17) and

10. When emphasizing that the event in the consequent occurs/occurred at a specific time or within a specific time interval, it is more appropriate to use a clause containing *toki*. Compare the following examples.

(a)

Yuugata ni-naru-to	kaze	ga	tsumetaku-nat-ta.
evening become-COND	wind	NOM	cold-become-PAST

"The wind became cold in the evening."

(b)

Yuugata ni-nat-ta	toki	kaze	ga	tsumetaku-nat-ta.
evening become-PST	when	wind	NOM	cold-become-PAST

"When it became evening, the wind got cold."

The difference between these sentences is that while the former expresses the natural temporal connection between it becoming evening and the wind getting colder, the latter emphasizes the point at which evening came and at which it became colder. A *toki* clause emphasizes the completion of the event in the subordinate clause as expressed by the past tense *ta* form. The *to* clause simply indicates a fact.

(18) express situations observed by a speaker, in which time is not specifically emphasized in either the antecedent or the consequent. *To* sentences that express past events tend to lose their emphasis on contingency relationships, instead emphasizing the sequentiality of events expressed in the antecedent and consequent, observing the whole event objectively and externally. For this reason *to* sentences are often used as descriptive expressions in novels, even when the agent is the speaker him/herself. An activity that was performed by the speaker him/herself can be viewed objectively and externally. Please note the following examples:

(19) Kare ni wakare o tsugeru-to kossori namida o nugut-ta.
he DAT farewell ACC tell-COND secretly tear ACC wipe-PST
"When I bade him farewell I secretly wiped my tears."

(20) Watashi ga soo iu-to kare wa makka ni-nat-te okot-ta.
I NOM so say-COND he TOP red-become-GER angry-PST
"When I said so, he got very angry."

(21) Fuyu ni-naru-to haha to yoku Hawai e it-ta.
winter become-COND mother COM often Hawaii LOC go-PST
"I often went to Hawaii with my mother in winter."

Sentences (19) and (20) describe past one-time events, while sentence (21) describes a past habitual event. Each describes an event externally and objectively and sounds as though the speaker is impersonally reporting the events.[11] For this reason, when a *to* sentence describes the speaker's own act, the sentence becomes rather awkward as the speaker is observing his/her own act as a third person. Such is evident, for example, in Hasunuma (1993): *Yuube gohan o taberu-to terebi o mimashi-*

11. Hasunuma (1993) also describes the function of *to,* concluding that nonhypothetical *to* is used when a speaker views the situation externally as an observer.

ta "When I finished eating dinner last night, I watched TV." The awkwardness is derived not merely from the speaker's volitional controllability of the event in the consequent, but also from the external observation of the event that a *to* sentence fundamentally depicts.

This observation that the function of *to* is to describe a past event externally and objectively somewhat communicates the notion of "discovery" proposed by Toyota (1979), who states that when a *to* sentence is used to express a one-time event and various other conditions are satisfied, the *to* sentence can express the meaning of "discovery." Although such interpretation of *to* sentences tends to depend on the combination of predicates in the antecedent and consequent, and in the absence of a unified theory, for certain predicates, it is reasonable to say that *to* sentences do, at least in part, express "discovery." This function, however, is not exactly the same as the function of *tara* in past context (discussed in section 5.2.3.1). The use of *to* sentences in past context is to describe a whole sentence as an objectively perceived situation. Thus far, it can be concluded that *to* sentences are used to express habitual/general (whenever) and one-time (when) events in past context.

5.1.2.2 Hypothetical To

The above meanings expressed by *to* in past context are reasonable, since each refers to observable, historically perceptible events. In order to see whether *to* sentences can express a hypothetical meaning in past context, sentences (13)–(16) and others will be examined in past context. Note that the tense of the predicate in the consequent for sentences (13)–(16) is the only object of alteration. For example:

(22)	Kinoo	hareru-to	yokat-ta	naa.
	yesterday	sunny-COND	good-PST	PART

"It would have been nice if it was sunny yesterday (I wish it was sunny yesterday)."

(23)	Uchi	no	musuko	wa,	moo	sukoshi	hansamu-da-to
	family	GEN	son	TOP	more	little	handsome-COP-COND

moteta-n-da	kedo	nee . . .
popular-PST-COMP-COP	but	PART

"As for my son, if he were a little more handsome, he would have been popular among girls."

(24)	Tegami	de	koogi suru-to	yokat-ta	ne.
	letter	INS	protest-COND	good-PST	PART

"It would have been good if you had protested by sending a letter./You should have protested by sending a letter."

(25)	Tomu Hankusu	ga	shuyaku	o	enjiru-to
	Tom Hanks	NOM	leading part	ACC	play-COND

eiga	wa	omoshirokat-ta.
movie	TOP	interesting-PST

"When/whenever Tom Hanks played the leading part, the movie was interesting."

(26)	Anata	ga	ik-anai-to	minna	ga	komat-ta.
	you	NOM	go-NEG-COND	everyone	NOM	trouble-PST

"When/whenever you did not go, people were troubled."

(27)	Kane	o	das-anai-to	shijitsu	o	barashi-ta.
	money	ACC	give-NEG-COND	truth	ACC	expose-PST

"When/whenever he did not give me the money, I exposed the truth."

It is intriguing that sentences (22)–(24) clearly express counterfactual meanings, whereas it is almost impossible to identify such in sentences (25)–(27). The latter sentences, which involve a stative predicate or nonstative predicate in both the antecedent and the consequent, most likely receive either a "whenever" or a "when" interpretation only. This observation indicates that human volitional controllability (present or nonpresent in the predicates) does not seem to be the key to determining the meaning. The hypothetical interpretation seems to be created only by the use of *yokat-ta* as a predicate in the consequent and some other devices, such as *kedo* "but," *noni* "though," and so forth, that create a counterfactual meaning. In fact, *-to yokat-ta* means "I wish" and is considered a counterfactual set phrase. The chances that *to* sentences receive a hypothetical interpretation in past context is therefore slim due to the fact that the use of conditional *to* is primarily related to events in the actual world domain. Even though stative predicates, for example, increase the degree of hypotheticality (since they create a world that is spatiotemporally distant from the speaker, especially in combination with future context, i.e., nonactual world domain), such behavior of stative predicates is expunged in past context (the actual world domain). Hence, a grammatical device is required in order for *to* sentences to express a hypothetical meaning in past context.

Although future events are essentially hypothetical, some events are considered known to be true due to human insight and the ability to observe situations objectively, and *to* represents these nonhypothetical worlds. For worlds in which a speaker's volition is insufficient to control and manipulate a situation, however, *to* is not sufficiently vigorous to render such events predictable. This is the reason why *to* can be used to express hypothetical situations. This feature, however, becomes less distinct when *to* is used in past context. Only some cases with certain devices, such as the use of *yokat-ta* "I wish" and *kedo* "though," can rectify the hypothetical meaning. The concept of "past" is perhaps strong enough to entice the minor function of conditional *to,* since past events are after all realized, historicalized events where there is no longer any room to hypothesize the consequence of events. Occasions where *to* sentences are used to express a hypothetical (counterfactual) meaning are therefore relatively less frequent compared to other functions of *to* sentences. Nevertheless, it is certainly true that *to* is used to express hypothetical situations in past context as well.

5.1.3 Summary of the *To* Conditional

Observing *to* sentences in both nonpast and past context, it has become clear that *to* sentences can express not only "whenever" and "when" (nonhypothetical) meanings, but also an "if" (hypothetical) meaning. Nevertheless, if the meanings of *to* are divided equally among these three, then the hypothetical meaning occupies only a third of all possible functions; the majority (two-thirds) are nonhypothetical. Therefore, the ratio in which *to* sentences express modality is relatively low.

Thus, *to* sentences are predominantly used to express nonhypotheticality. This observation is derived from the fact that unpredictable or indeterminable situations cannot co-occur in the consequent of *to* sentences, and that predictable and general events are expressed by *to* sentences. Nonhypothetical or hypothetical interpretations of these two conditional forms are related to the meaning of the forms: imperfective forms expressing generality and habituality, and perfective forms expressing specificity. Generality, habituality, and specificity can be approached either temporally or from an experiential viewpoint. When looking at these three notions from an experiential viewpoint, generality and habituality are produced by iterated experience, but this is not true for specificity. When considering the notions from a temporal viewpoint, it is easy to predict consequences in events that express generality and habituality, whereas, again, such is not the case with specificity. Either view leads to the interpretation that generality and habituality invite nonhypothetical interpretations, and specificity invites a hypothetical interpretation.

In the English study of conditionals, Dancygier (1998) also examined the patterns of verb forms[12] used in conditional sentences and suggested that the function of prediction is central to the definition of conditionality, and that conditionals display certain formal features, such as verb forms, typical clause order, or intonation. The degree of hypotheticality expressed by a conditional form is certainly correlated with verb forms, as observed in conditional sentences in Japanese. Some researchers (e.g., Soga 1983) claim that nonpast tense forms are grammatically required in *to* clauses. However, the examination of the features of *to* sentences has revealed that nonpast tense forms predominantly expressing habitual or future events are not accidental phenomena in the Japanese language; rather, the nonpast tense form plays an active role in deciding the meaning of the *to* antecedent.

5.2 HYPOTHETICALITY AND THE *TARA* CONDITIONAL

Imperfective forms are used to express habituality, and habituality is concerned with nonhypothetical situations. This feature of imperfective forms is related to their use in *to* sentences, and although the use of such fully depends on a speaker's knowledge and perception, it is precisely the *to* form that is most often used to express nonhypothet-

12. Examples are the following: "If it rains/rained/had rained, the match will be/would be/would have been cancelled."

icality. What, then, about perfective forms? Kuno (1973) stated that "semantically, *tara* in 'S1 *tara* S2' always presupposes that the action or state represented by the verb it is attached to ' is completed or finished before' S2 takes place" (p. 177), which indicates the sequential feature of *tara* sentences. As described below, *tara* is viewed as possessing bipolar functions, expressing both hypothetical and nonhypothetical worlds. What factors, then, determine whether *tara* sentences invite hypothetical or nonhypothetical interpretations? This section clarifies the way that perfective forms are used in conditional sentences in relation to hypotheticality.

The conditional form *tara* consists of the past tense suffix *ta* and the mora *ra*. The past tense suffix *ta* has been viewed as perfective in the traditional study of Japanese tense and aspect. This perfectivity led to opposing views of time: perfectivity as punctual vs. imperfectivity as durative. The use of the past tense suffix *ta* in a simple sentence indicates the realization of a past event, whereas the use of *ta* in a subordinate clause indicates the realization of an event relative to an event in a main clause. Such realization of an event can be viewed as pointing to one-time events as opposed to general and/or habitual events, an aspectual feature that affects the interpretation of *tara* conditional sentences as hypothetical or nonhypothetical.

5.2.1 The Temporal Character of *Tara*

As observed in the previous section concerning the conditional *to*, the more that an experience is iterated, the more one internalizes the experience as a natural course of events; that is, the consequence of an event becomes predictable. Internalized events receive nonhypothetical interpretations in a conditional sentence, including a sentence in which an event is, on the contrary, a temporary experience; then, a speaker cannot make a persuasive prediction regarding the situation in question. A speaker may, however, hypothesize the realization of the situation, and since *ta* can be viewed as expressing an event punctually, it can be surmised that *tara* sentences express the hypotheticality of a future event.

In order to illustrate this point, let us first consider some examples that compare the generality and habituality of *to* with the temporality of *tara*:

(28) a.

Tenki	ga	yokat-tara	tsuri	ni	ku.
weather	NOM	good-COND	fishing	DAT	go

"If the weather is good tomorrow, I will go fishing."

b.

Tenki	ga	ii-to	tsuri	ni	iku.
weather	NOM	good-COND	fishing	DAT	go

"When the weather is good I (always) go fishing."

While sentence (28)b most likely receives a habitual interpretation, that is, the speaker always goes fishing when the weather permits, (28)a does not give the impression that the speaker's attitude is habitual, but rather only refers to a specific occasion. The notion of specificity expressed by *tara* can further be identified in the following sentences that accompany temporal adverbial phrases:

(29) a.

Ashita	tenki	ga	yokat-tara	tsuri	ni	ku.
tomorrow	weather	NOM	good-COND	fishing	DAT	go

"If the weather is good tomorrow, I will go fishing."

* b.

Ashita	tenki	ga	ii-to	tsuri	ni	ku.
tomorrow	weather	NOM	good-COND	fishing	DAT	go

"If the weather is good tomorrow, I will go fishing."

?? c.

Tenki	ga	yokat-tara	itsumo	tsuri	ni	iku.
weather	NOM	good-COND	always	fishing	DAT	go

"If the weather is good I always go fishing."

d.

Tenki	ga	ii-to	itsumo	tsuri	ni	iku.
weather	NOM	good-COND	always	fishing	DAT	go

"If the weather is good I always go fishing."

The above examples exhibit the contrastive function of *to* and *tara*. While *tara* can co-occur with a phrase that expresses a one-time event as seen in sentence (29)a, the addition of the adverb *itsumo* "always" makes the *tara* sentence awkward, as seen in sentence (29)c. The fact that the generality of time expressed by *itsumo* is not compatible with *tara* supports the idea that *tara* is concerned with specific events. This observation can also be made when considering *tara* sentences that refer to past events, as exemplified in the following sentences. Note once again that the English translation does not accurately reflect the structural problems of the original Japanese:

(30) a. Kinoo kare ni at-ta toki, nihongo o hanashi-tara
yesterday he DAT meet-PST when Japanese ACC speak-COND
totemo home-rare-ta.
very admire-PASS-PST
"When I went to Japan last year, I was very admired when I spoke Japanese."

?? b. Kinoo kare ni at-ta toki, nihongo o hanasu-to
yesterday he DAT meet-PST when Japanese ACC speak-COND
totemo home-rare-ta.
very admire-PASS-PST
"When I went to Japan last year, I was very admired when I spoke Japanese."

The *tara* in sentence (30)a can co-occur with a subordinate clause that indicates a specific point in time, such as "when I met him yesterday," whereas the *to* construction in (30)b is awkward when this type of temporal subordinate clause is employed. Without the subordinate clause, both sentences are acceptable, except that the *tara* in (30)a expresses the unpredictability of being admired at the time the speaker was in Japan, while the *to* in (30)b does not create such an implication.[13]

13. This point will be further examined later in this section. For a detailed discussion, please refer to example sentences (39)–(42), which exemplify *tara* sentences involving past events in their consequents.

Thus all these sentences support the notion that *tara* sentences do not express the generality and habituality that communicate nonhypotheticality, but the aspectual feature of *tara* can be viewed in terms of the notion of specificity. The more frequently an event occurs, the greater the degree of generality and habituality, since the speaker can predict the causal relation of the event. Specific events such as one-time events, on the other hand, cannot exhibit such character. In the diagram below, each dot represents a single occurrence of the same event (this of course is actually a pair of events—the antecedent and the consequent such as *ame ga* furu, it rains," and *shiai wa chuushi-da*, "the game is/will be cancelled"):[14]

(31)

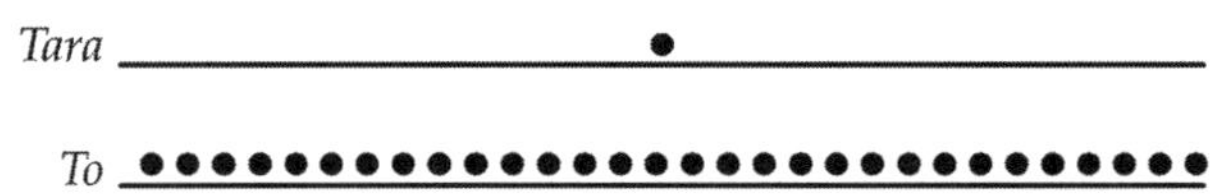

As the diagram illustrates, frequently occurring events are described by *to* sentences. On the contrary, *tara* sentences indicate a one-time event that can be approached from two standpoints: frequency of experience (a one-time event rather than the iterated experience necessary to predict a causal relation); and contingency relationships of two events expressed in the antecedent and the consequent of a *tara* sentence. The latter communicates the sequentiality expressed by *to* sentences, such as that which is identified in example sentences (28)a and (28)b, although these two sentences do not share exactly the same meaning due to the difference in frequency of occurrence. This difference will be further clarified in the following discussion.

Thus, while *to* sentences are mainly used to express nonhypotheticality due to the predictability gained through iterated experience, *tara* sentences are used to express a one-time event the truth of which may or may not be predictable to the speaker. The ability of *tara* to express hypotheticality is thus attributed to the aspectual character

14. The exact number of dots here is irrelevant since experiences vary depending on the individual as well as the type of event. Past, present, and future demarcations are also irrelevant since all are equally applicable.

that possessed by *ta*. The lack of predictability caused by the lack of frequency in experience communicates the notion of indeterminability, which is fundamental to the notion of hypotheticality. *Tara* can thus be used to express hypothetical situations where a speaker cannot determine the truth or realization of the antecedent. Situations that are indeterminable to a speaker are captured in a nonpast context, since past events can no longer be considered as hypothetical unless they express counterfactual events.

5.2.2 *Tara* Sentences in Nonpast Context

5.2.2.1 Hypothetical Tara

Of course, whether a situation is determinable or indeterminable is based on the speaker's knowledge and perception about the world. If the antecedent is concerned with a scientific/mathematical fact or a commonly understood fact, the antecedent inevitably receives a nonhypothetical interpretation. As was discussed earlier, the notion of indeterminability is often associated with situations in which a speaker does not have sufficient experience in predicting the truth or realization of events. Specific events described as one-time events with *tara*, therefore, tend to receive a hypothetical interpretation. Also, the notion of indeterminability communicates at its root that the truth or realization of a situation is beyond the speaker's volitional control.

As one-time events occurring with *tara*, natural phenomena, another person's activities, events described by stative predicates, and so forth are also likely to receive a hypothetical interpretation in *tara* sentences. Natural phenomena, such as weather and earthquakes, are fundamentally unpredictable. Similarly, the state of mind of another individual is not directly experiencable or recognizable; it is beyond a third person's volitional control and cannot be perceived as expressing generality unless the speaker has firm evidence that it is true or actualized. Events described by stative predicates are also beyond a speaker's volitional control. Therefore, this type of predicate normally invites a hypothetical interpretation. Below, sentence (32) concerns a natural phenomenon; sentences (33) and (34) involve a third party; and sentences (35) and (36) include stative predicates in the antecedent. In each of these *tara* sentences, an "if" interpretation is better received than a "when" interpretation:

(32) Ookina jishin ga dokanto ki-tara doo shi-yoo.
big earthquake NOM suddenly come-COND how do-VOL
Tookyoo wa ikkan no owari-da.
Tokyo TOP completely GEN end-COP
"If a big earthquake comes all of sudden, what shall I do? Tokyo will be completely destroyed."

(33) Minami ga iku-tte it-tara, ore mo iku.
Ms. Minami NOM go-QUOT say-COND I also go
"If Minami says that she would go, I will go, too."

(34) Kare ga yame-tara gakubu wa hookaisuru.
he NOM resign-COND department TOP collapse
"If he resigns, the department will collapse."

(35) Oishikat-tara taberu kedo, mazukat-tara tabe-nai yo.
delicious-COND eat but bad-COND eat-NEG VOC
"If it is delicious I will eat it, but if it is not, I will not eat."

(36) Atashi ga ik-anakat-tara aitsu mo ik-anai-tte yo.
I NOM go-NEG-COND he also go-NEG-QUOT VOC
"He says that he would not go if I do not go."

As was demonstrated in the examination of sentences (11) and (12), *to* conflicts with a volitional expression in the consequent since such does not communicate the notion of generality that *to* sentences primarily involve. *Tara,* however, does not exhibit such circumscription; hence, the *tara* construction can invite a volitional expression in the consequent, as exemplified in sentences (32), (33), and (35). Thus, the fundamentally indeterminable aspects described in sentences (32)–(35) invite a hypothetical interpretation, and the undetermined event in the consequent is compatible with the antecedent.

The characteristics of indeterminable situations should also include events described by unaccusative verbs, since such predicates are concerned with situations that are beyond a speaker's volitional control. This observation certainly contributes to the interpretation of a *tara* sentence, yet, it must be noted that the characteristics of unaccusativity alone—that is, a lack of speaker volitional control over the situation—is not sufficient for the interpretation of the *tara* construction with unaccusative verbs. This can be observed in the following examples:

(37)	Mondai	ga	kaiketsushi-tara	suguni	shirasete	kudasai.
	problem	NOM	resolve-COND	right away	inform	please

"If the problem gets resolved, I will inform you right away."

(38)	Mondai	o	kaiketsushi-tara	suguni	shirasemasu.
	problem	ACC	resolve-COND	right away	inform

"When I resolve the problem, I will inform you right away."

(39)	Honda	ga	ki-tara	boku	wa	kaeru.
	Mr. Honda	NOM	come-COND	I	TOP	go home.

"When/If Mr. Honda comes, I will go home."

Although the contribution of intransitivity can certainly be identified in the hypothetical interpretation of sentence (37) compared to the nonhypothetical interpretation of sentence (38), it is also true that the meanings are ambiguous from these sentences alone. For example, sentence (39) includes not only an unaccusative verb, but also the third person Mr. Honda; thus, the interpretation is absolutely context dependent. If the situation is such that the speaker has no idea regarding Mr. Honda's schedule, the interpretation should be "if." On the other hand, if the speaker is expecting Mr. Honda to come, the interpretation is "when." In the latter case, an appropriate interpretation would often be "after his arrival," which excludes the hypothetical connotation altogether. The same observation can be made from sentence (38), where either interpretation is possible depending on the speaker's perception.

5.2.2.2 Nonhypothetical Tara

Although the characteristics of unaccusative verbs carry indeterminability that is prone to render *tara* sentences hypothetical, it is also true that there are cases where either interpretation is possible or even where a nonhypothetical interpretation prevails. This nonhypothetical interpretation is not born from a speaker's repetitive experience that is registered as a predictable event, but is due to the speaker's knowledge, that is, the determinability of the situation viewed by the speaker. Even though the future is not perfectly predictable, and the situation may change afterward, the speaker's knowledge allows a *tara* sentence to receive nonhypothetical interpretation. Also, combining this nonhypothetical interpretation with the aspectual character of *ta* (perfectivity) often generates an "after" interpretation that emphasizes the sequentiality of two events in a *tara* sentence. Please note the following examples:

(40)

Kono	mondai	o	to-itara	uchi	e	kaeru.
this	problem	ACC	resolve-COND	home	LOC	return

"When/after I resolve this problem, I will go home."

(41) a.

Mukoo	ni	tsu-itara	tegami	o	kaku.
over there	LOC	arrive-COND	letter	ACC	write

"When/after I get there I will write a letter."

b.

Hima	ga	at-tara	tegami	o	kaku.
free time	NOM	have-COND	letter	ACC	write.

"If I have free time, I will write a letter."

(42)

Kono shigoto	ga	katazu-itara,	mina	kyuuka	o
toru	daroo.				
this job	NOM	finish-COND	everyone	vacation	ACC
take	AUX				

"When this job is done, everyone will take a vacation."

(Masuoka 1993)

All three examples are interpreted as expressing nonhypotheticality due to the determinability of the events viewed by the speaker. Jacobsen (1992a) compared sentence (41)a with sentence (41)b that includes a stative predicate in the antecedent and demonstrated that an achievement component of the meaning in the *tara* antecedent in sentence (41)a is prone to expressing low hypotheticality as compared to a stative predicate. An unaccusative verb such as *tsuku* in (41)a gives us a clear indication of exactly when the event is realized; in our mind's eye it can easily be visualized (experienced). Stative predicates, such as *aru* "something exist(s)," however, express a kind of temporal latitude as opposed to a specific point of realization, and although this observation is certainly valid, the achievement component of the meaning is but a partial determinant in the degree of nonhypotheticality. Masuoka (1993) also pointed out that when the event in the antecedent can be perceived as being realized within a certain time frame, then such an event contributes to the nonhypothetical interpretation of the sentence, as in example (42).

Although a slight difference in the interpretation might be recognized in terms of the degree of hypotheticality between sentences (41) and (42), the difference is attributable to the degree to which the speaker knows the realization/truth of the event. When the event is about the speaker himself, as in sentence (41), the interpretation is undoubtedly nonhypothetical due to the speaker's self-knowledge. Thus, even though some cases can be theorized in terms of the type of predicate (such as an unaccusative verb) and type of event that is beyond the speaker's volitional control (such as weather or a third person's activity), an interpretation of *tara* in future events is fundamentally based on the speaker's knowledge about the future events.

We have seen that *tara* sentences concerning a future event can express either hypotheticality or nonhypotheticality, with the former interpretation brought about by the indeterminability of a future event. A single experience does not normally provide sufficient evidence upon which to form an accurate generalization of an event, and therefore in a hypothetical situation, the causal relationship is less distinct and the antecedent is emphasized. The antecedent is presented as a condition meaning "provided that," and the antecedent is emphasized since it is the key to whether or not the consequent is realized. Nonhypothetical *tara* sentences, on the other hand, emphasize the sequence of events involved. This interpretation is brought about when a speaker is confident in the realization of a sequential event—particularly the case, for example, when the realization of the event is guaranteed by the speaker's self-knowledge, even though, depending on the type of predicate, there is a variation in the degree of hypotheticality. *Tara* is thus a resident of the speaker's domain, while *to* belongs to the domain of objectively perceptible worlds.

5.2.3 *Tara* Sentences in Past Context

5.2.3.1 Nonhypothetical Tara

Tara, which expresses a one-time event generated by its aspectual character of perfectivity viewed as punctual, is interpreted as first-time recognition (experience) of an event when the sentence receives a nonhypothetical interpretation in past context. The notion of first-time recognition at root communicates the notion of "discovery," which also indicates that the situation expressed in the causal relation was, from the speaker's viewpoint, unexpected. *Tara* is thus used to express "discovery," "surprise," or "unexpectedness" when used in a past context. Consider the following examples:

(43)

Kare	wa	heya	ni	hait-tara	uwagi	o	nui-da.
he	TOP	room	LOC	enter-COND	jacket	ACC	take-off-PST

"When he entered the room he took off his jacket."

(Koide, Komatsu, and Saita 1981)

(44)

Imooto	wa	atarashii	apaato	ni	hikkosh-itara
my sister	TOP	new	apartment	LOC	move-COND

mae yori akaruku-nat-te, tomodachi mo takusan tsukut-ta.

before than cheerful-become-GER friend too many make-PST

"When my sister moved into a new apartment, she became more cheerful than before and made many friends."

(45) Keeki o tabe-tara kyuuni onaka ga itaku-nat-ta.
cake ACC eat-COND suddenly stomach NOM hurt-become-PST
"When I had a piece of cake, I suddenly got a stomach ache."

(46) Hontoo no koto o it-tara atama o nagu-rare-ta.
truth GEN thing ACC say-COND head ACC hit-PASS-PST
"When I told the truth I was hit on the head."

Sentence (43) was considered inappropriate by Koide, Komatsu, and Saita (1981) and Jacobsen (1992a) due to the fact that the agent (some third person) is common to both the antecedent and consequent. They noted, however, that the accompanying *soto de wa nuganakat-ta noni* "although he did not take off his jacket outside" makes the sentence appropriate. Indeed, sentence (43) is not inappropriate,[15] since *tara* implies that the speaker is astonished by the fact that, on a hot summer day for example, the man took off his jacket inside a chilly, air-conditioned room, but left it on while he was outside. Koide's interpretation may have been a result of the descriptive style of the sentence itself, and by the fact that *to* would have been a more obvious choice so as to avoid the implication created by *tara*. Sentence (44) supports the appropriateness of (43), since the latter can be accepted without hesitation, even though the agent is common to both the antecedent and the consequent. The implication here is that the agent was somewhat unsociable in her former apartment building, so the speaker finds the fact that she became much more cheerful surprising. The same can be observed from sentences (45) and (46). *Tara* used in past context thus connotes an unexpectedness felt by the

15. This is not the case, however, for a future event that does not involve the speaker, because a sense of discovery is not applicable in an unrealized situation. A speaker cannot have been surprised by an event if that event has not occurred.

* Mondai ga kaiketsu shi-tara Ritsuko wa hitoride ryokoo suru.
problem NOMresolve-COND Ritsuko TOP alone trip"
Ritsuko will go on a trip alone when the problem gets resolved."

speaker, and therefore the events described by a *tara* sentence psychologically are much closer to the speaker him/herself.

The situation viewed and felt by the speaker is the key difference between *tara* and *to* regarding use of the latter *to* express the notion of "discovery" as proposed by Toyota (1979). As mentioned earlier, the notion of "discovery" communicates "surprise" or "unexpectedness" brought about in the speaker, and the use of *tara* in past context can be considered a device by which the speaker expresses how s/he felt about the situation without using overt phrases such as *odoroi-ta* "I was surprised" and *Soo wa omow-anakat-ta-n-da kedo . . .* "I did not expect that to happen, but . . . ." *Tara* sentences describe the situation viewed by the speaker internally. On the other hand, the use of *to* is to describe the situation externally and objectively. In this sense, *tara* exhibits a psychological closeness to the speaker, whereas *to* is relatively distant. Hasunuma (1993) mentioned that sentences like *Taroo ga ki-tara, Hanako ga kaet-ta*, "When Taroo came, Hanako left," [presented by Kuno (1973)] are rarely seen in written documents. Since this sentence subtly expresses the speaker's feelings regarding the situation—that is, that the speaker was surprised to see Hanako leave upon Taro's arrival—it is not surprising that this type of sentence is seldom used in descriptive written documents. Instead, *to* is used.

The above argument explains why *tara* sentences that refer to the past do not accept consequents that are expressions of self-knowledge. Self-knowledge contradicts the sense of "discovery" and "surprise" carried by *tara*. The awkwardness of two of the following examples supports this observation:

(47) *	Mario	ga	ki-tara	watashi	wa	kaet-ta.
	Mario	NOM	come-COND	I	TOP	leave-PST

"When Mario came I left.

(48)	Mario	ga	ki-tara	Kanako	wa	kaet-ta.
	Mario	NOM	come-COND	Kanako	TOP	leave-PST

"When Mario came Kanako left."

(49) *	Nihon	ni	it-tara	unagi	ga	tabe-takat-ta.
	Japan	LOC	go-COND	eel	NOM	eat-DES-PST

"When I went to Japan, I wanted to eat eel."

Sentences (47) and (49) are awkward due to the irreconcilability of the speaker's volition or desire with a sense of unexpectedness.[16] For example, the awkwardness of sentence (47) is attributable to a contradiction between the speaker's will, *kaet-ta* "left," and the unexpectedness, discovery, or surprise of the event. Since the speaker controls his/her own volitional activity, the speaker's leaving is predictable from the his/her viewpoint and should not lead the speaker to react with surprise. If the agent of the consequent in (47) was another individual, then the sentence would be appropriate since another individual's activities are not experiencable or controllable, that is, they are not necessarily predictable. The appropriateness of sentence (48) can be attributed to this observation. Thus the use of *tara* in past contexts is highly "speaker-oriented."

5.2.3.2 Hypothetical Tara

The notion of the one-time event created by *tara* communicates the idea of first-time recognition of the event, and this recognition further communicates the notion of "discovery" and/or "surprise." Although most *tara* sentences create such connotations in everyday life communications, this observation does not imply that *tara* in past context is never viewed as expressing hypotheticality. Although it is true that *tara* cannot be used to express past habitual events, it can be used to express hypotheticality through counterfactual interpretation. Please note the following examples:

(50) a. Mainichi renshuu shi-tara motto umaku-nat-ta.
everyday practice-COND more good-become-PST
"When I practiced every day, I was surprised to find myself improving."

b. Mainichi renshuu shite-i-tara motto umaku-nat-ta.
everyday practice-PERF-COND more good-become-PST
"If I (you) had practiced every day, I (you) would have seen more improvement."

16. These sentences can be perceived as appropriate only when the speaker does not have self-knowledge and is surprised by his/her own action. In such a case, the interpretation of sentence (47) would be "I was surprised to see myself leaving when Mario came. I'm not sure, but I guess I do not like to be with him."

c. Mainichi renshuu shi-tara motto umaku-nat-ta noni nee . . .

everyday practice-COND more good-become-PST though PART

"If I (you) practiced every day, I (you) would have seen more improvement."

d. Mainichi renshuu shite-i-tara motto umaku-natte-i-ta.

everyday practice-PERF-COND more good-become-PERF-PST

"If I had practiced every day, I would have improved more."

(51) a. Uso o tsu-itara atama o nagu-rare-ta.

lie ACC say-COND head ACC hit-PASS-PST

"When I told a lie, I was hit on the head."

b. Uso o tsu-itara atama o nagu-rare-ta daroo . . .

lie ACC say-COND head ACC hit-PASS-PST AUX

"If I had told a lie, I would have been hit on the head."

c. Uso o tsuite-i-tara atama o nagu-rare-ta.

lie ACC say-PERF-COND head ACC hit-PASS-PST

"If I had told a lie, I would have been hit on the head."

d. Uso o tsuite-i-tara atama o nagu-rarete-i-ta.

lie ACC say-PERF-COND head ACC hit-PASS-PERF-PST

"If I had told a lie, I would have been hit on the head."

Sentence (50)a is semantically awkward since it expresses the speaker's sense of surprise in seeing an improvement that resulted from everyday practice. In other words, sentence (50)a is presenting the speaker's disbelief and astonishment (contrary to common sense) that if one practices everyday s/he improves. The speaker's nonsensical pragmatism leads us to a hypothetical interpretation, namely, that the speaker had not practiced everyday and the consequence is that the speaker has made no progress. In order to explicate such a hypothetical meaning, however, some grammatical devices, such as the addition of *noni,* "though," which implies the opposite/different result, the use of *te-iru* that makes the predicate perfective, and so forth, are required. These devices are exemplified in sentences (50)b–(50)d. Of the three sentences, (50)d, which uses *te-iru* in both the antecedent and the consequent expresses the highest counterfactuality and therefore carries a strong "invited inference" that the actual event happened in the past.

The same observation can be made of sentences (51)a–(51)d. The connotation of sentence (51)a is that the speaker was surprised by the fact that s/he was hit on the head, even though such an act may not be considered inconceivably unreasonable when someone tells a lie. This sentence is presented to show the intermingling of a past event with the speaker's feelings. Sentences (51)b–(51)d, on the other hand, express a counterfactual meaning and, again, some devices are accommodated. Without these devices and in the absence of pragmatic or sociocultural justification to make the sentence sensical, *tara* expresses a past one-time event that has already occurred.

5.2.4 Summary of the *Tara* Conditional

Tara conditional sentences are approached from the viewpoint that the hypothetical meaning expressed can be attributed to the perfective interpretation of the past-tense suffix *ta,* which is viewed as punctual, that is, it expresses a one-time event. This interpretation allows exploration of various functions of *tara* sentences in that they are used to express nonhypotheticality and hypotheticality in both future and past contexts.

In nonpast context, the nonhypothetical interpretation is generated by the speaker's knowledge about the world: if the truth of the event described in the antecedent is known to the speaker, then the sentence receives a nonhypothetical interpretation (when). *Tara,* however, does not receive a "whenever" interpretation since the notion of a "one-time event" cannot communicate the notion of habituality. A hypothetical interpretation, on the other hand, generally originates in certain factors, such as a third person's activity, natural phenomena, and/or events described by unaccusatives. These events, at root, all communicate the notion of indeterminability and uncontrollability of the event's realization viewed by a speaker. When an event is beyond the speaker's volitional control, a *tara* sentence expresses a hypothetical meaning (if). Also, the specificity of *tara* regarding the occurrence of an event as opposed to habituality creates an interpretation that the consequences of two events expressed in the antecedent and the consequent make the speaker perceive the realization of an event unpredictable.

In past context, the specificity fundamentally carried by *ta* communicates the concept of "first time recognition," which leads *tara* sentences to receive the meaning of "discovery," "surprise," and/or "unexpectedness." Since these notions contradict the speaker's own knowledge and *tara* sentences are therefore awkward when the speaker him/herself is the subject of the antecedent and the consequent, most sentences can be interpreted as expressing the speaker's feeling of unexpectedness.

Tara sentences thus express a speaker's observation internally, and therefore may rarely be seen in a descriptive written document. *To,* which is used to express a situation externally and objectively viewed by a speaker, is used instead. In order for a *tara* sentence to receive a hypothetical interpretation in past context, grammatical devices, such as *noni*, which express a counter situation, are required. Therefore, taking both contexts into consideration, the degree to which *tara* sentences express a nonhypothetical meaning is greater than the degree to which they express a hypothetical meaning.

5.3 HYPOTHETICALITY AND THE *BA* CONDITIONAL

In past studies, *ba* has been variously described as: expressing a connection between a condition and a causal relation (Alfonso 1966, Koide, Komatsu, and Saita 1981); implying an unmentioned opposite alternative in the antecedent (Hinds and Tawa 1976); expressing a speaker's hopes or wishes that a situation will be true (McGloin 1976); expressing hypotheticality because of the grammatical requirement that it not be used in a past context (Koide, Komatsu, and Saita 1981); and expressing a general causal relation that is established beyond the time framework depending on the combination of the antecedent and the consequent (Masuoka 1993). *Ba* certainly has all of these functions, yet identifying particular semantic functions is only a part of the process of defining the fundamental function of the *ba* conditional. *Ba* sentences herein are approached from the viewpoint of modality, and it is found that in terms of hypotheticality and nonhypotheticality, *ba* can in fact be used to express both, although in day-to-day conversation, it most often expresses the former. The question is, then, what factors result in the expression of nonhypotheticality or hypotheticality in *ba* sentences and to what degree. The formation of *ba* constructions is such that *ba* follows the so-called "e" form of a predicate, for example, *kak-eba* "if you write," and *omoshiroke-reba* "if it is interesting." The *"e"* form of a predicate does not possess independent meanings or functions the way a *ta* form used in conjunction with *tara* does or a *ru* form preceding a *to* form does. It seems that *ba* sentences emphasize neither a one-time event (as with *tara* sentences) nor a natural course of events (as with *to* sentences). The former observation is especially true when *ba* is employed in reference to realized situations (actual world situations), where it refers to general and habitual events, but does not indicate a single specific event. The following cases, where a one-time event is about to be realized, further exemplify this observation:

(52) * a. Abunai! Otos-eba wareru!!
watch out drop-COND break
"Watch out! If you drop it, it will break!!" (McGloin 1976)

b. Abunai! Otoshi-tara wareru!!
watch out drop-COND break
"Watch out! When you drop it, it will break!!"

c. Abunai! Otosu-to wareru!!
watch out drop-COND break
"Watch out! When you drop it, it will break!!"

In sentences (52)a, b, and c, the event "to drop" appears about to be realized and the speaker shouts a warning. The immediacy of the event is indicated by the word *abunai*, "watch out." McGloin (1976b) pointed out that *ba* sentences are used to express a speaker's hopes or wishes that a given situation will be true, but in contrast *ba* cannot be used when a speaker intends a "warning." McGloin's statement is valid only as far as the immediacy of the warning is concerned. For example, *Ii? Kono kusuri o mainichi chanto nom-ana-kere-ba yoku-nar-anai-n-da yo* "Listen. If you do not take this medicine everyday regularly, you will not get better," is also a warning, but is an example where *ba* can be used. The difference is clearly related to a lack of immediacy. In the example of taking medicine, the speaker believes that the interlocutor has time to process the information before reaching a point where s/he must carry out the action described in the warning. Furthermore, such a warning is not limited to an exclusive occasion. In the warning in sentence (52)a, the speaker believes the immediacy of the situation is such that the interlocutor must take action immediately. The nonuse of *ba* is not due to the element of warning, but rather to the immediacy of that warning.[17] The fact that *ba* is not used to give such a warning supports the idea that *ba* is not an appropriate option to describe a single specific event in the domain of actual worlds, that is, an "(at the time) when" interpretation is not applicable.

The inappropriateness of the use of *ba* for a single, specific event can, however, be remedied by removing the word *abunai*, "watch out," as shown in the following example:

(53) Otos-eba wareru sa. Sonna koto atarimae daroo.
drop-COND break PART such thing natural COP
"(Glasses) break upon being dropped. It is natural, isn't it?"

17. The premise of the hypothetical nonuse and nonhypothetical use of *ba* can be further illustrated by examining the seemingly subtle differences between and "immediate warning" described above and a "threat or barter." Both *ba* and *tara* can be used in a causal relation that indicates a "threat," since the premise of a threat is that in the antecedent the speaker is expressing an indeterminable situation. In the following examples, there is no guarantee of the antecedent, which expresses a nonactual world situation, being realized.

(a) Kane o das-eba himitsu wa mamoru.
money ACC give-COND secret CONT keep
"If you give me money, I will keep the secret."
(Koide, Komatsu, and Saita 1981)

(b) Kane o dash-itara himitsu wa mamoru.
money ACC give-COND secret CONT keep
"If/when you give me money, I will keep the secret."

The realization of the antecedent in a "threat" sentence like this is not predictable or determinable, yet the consequent keeping of the secret is valid only on the realization of the antecedent.

Although the word *abunai* "watch out" could be used to indicate that the situation is about to be actualized, that is, that a situation is teetering on the edge of realization, once this word is removed, sentence (53) can be interpreted as expressing general knowledge expressed by the speaker. Such general knowledge about the world is acquired by iterated observation of previous events which in turn come to be accepted as general. Events iterated in the actual world, therefore, no longer belong in the domain of nonactual worlds, and in such a case the whole sentence acts as a hypothetical sponge, absorbing a multiple subject and expressing habituality and generality. This use of *ba* is often seen in a sentence with an "everyone" subject, which is why *ba* sentences are often interpreted as expressing a notion that is equivalent to "whenever" or "every time" in English, as in sentence (53). This again demonstrates that *ba* conditional sentences can be used to express not only hypothetical situations, but also nonhypothetical situations. Such nonhypothetical situations, however, cannot correspond to a one-time event as in *tara* sentences, but to events which a speaker can recognize as repeatable. In other words, *ba* sentences are used to express either a not-yet-realized situation, an indeterminable situation, or an iterated, general event.

A question arises, however, as to the difference between *ba* and *to*, which is also used to express a general, natural course of events. The following sentences are examples of some scientific facts that compare the use of *ba* and *to* (in which *to* is appropriately used):

(54) *a. Taiyoo wa higashi kara dere-ba nishi e shizumu.
sun TOP east from rise-COND west LOC set
"If the sun rises from the east, it sets in the west."

b. Taiyoo wa higashi kara deru-to nishi e shizumu.
sun TOP east from rise-COND west LOC set
"The sun rises from the east and sets in the west."

(55) a.	Haru-ni-nar-eba	sakura	ga	saku.
	spring-become-COND	cherry flowers	NOM	bloom
	"Whenever spring comes, cherry flowers bloom."			
b.	Haru-ni-naru-to	sakura	ga	saku.
	spring-become-COND	cherry flowers	NOM	bloom
	"When spring comes, cherry flowers bloom."			

Most Japanese speakers would consider sentence (54)a to be unacceptable or awkward, whereas sentence (55)a is acceptable and, in fact, frequently used. Both events—the sun's movement and the cherry flower bloom—are considered scientific facts that are recognized as expressing natural phenomena known by everyone and are taken for granted; neither sentence can be perceived as expressing hypotheticality. This is due to the fact that scientific facts exclusively exist in an actual world domain, and can no longer be viewed from an indeterminable world domain once they are established as facts—that is, the "if" interpretation is not applicable. As seen in sentences (54)b and (55)b, there is no problem in describing these phenomena with *to*, but this is not the case for *ba*. A question arises, then, as to what accounts for the appropriateness of sentence (55)a regardless of the fact that both events are exclusive to the actual world domain.

The key can be found in McGloin's research (1976) where she states that *ba* sentences express a speaker's hopes or wishes that a situation will be true. For most *ba* sentences, this claim is applicable as demonstrated in sentence (55) in which the speaker hopes for the arrival of spring and for the blossom of cherry flower. Sentence (55)b, on the other hand, does not create such connotation; it merely states the situation as a factual event as viewed by the speaker. Nevertheless, such a function of *ba* is not limited to positive connotation, but also includes adverse connotations, such as warnings, as exemplified in *Kono kusuri o nom-anakere-ba yoku-nar-anai*, "If you do not take this medicine, you do not get better." In this sentence, the speaker certainly does not hope the situation described in the sentence "do not take this medicine; do not get better" will be true, but the speaker's hope is expressed in the "invited inference" interpretation—"If you take this medicine, you will get better." McGloin's postulation should be slightly modified to reflect the fact that *ba* sentences express a speaker's expectation/perspective that the whole statement is valid from the speaker's viewpoint.

The inappropriateness of *ba* in sentences that express scientific fact seems due to this pragmatic function of *ba*. This is especially true for a sentence such as (54)a, which describes the sun's movement, and is seldom employed in everyday conversation, since we take such phenomena for granted. It is perhaps even rarer for one to look forward to or anticipate such an event as the sun's movement. Sentence (55)a, on the other hand, can easily be perceived as expressing the situation where the speaker is anxiously waiting for spring to come and for flowers to bloom. Therefore, it can be said that the appropriateness or inappropriateness of this type of sentence is attributed to singling out the sentence without providing any context. In fact, the addition of phrases such as *atarimae-da* and *futsuu-da* "it is naturally the case that ~," makes sentence (54)a as equally appropriate as sentence (55)a and can be accepted as naturally as sentence (53). This premise is applicable to most universal truths, scientific facts, and general courses of events.

The reason why *ba* is awkward in expressing these concepts alone, however, does not seem due solely to an insufficient supply of context, but also to the fundamental function of *ba*. That is, *ba* can vacillate between hypothetical and nonhypothetical worlds depending on the speaker's perception and context. For example, as mentioned above, scientific facts exclusively exist in an actual world domain, and they can no longer be viewed from a hypothetical world domain once established as fact. Yet, when *ba* is employed to express a scientific fact, it gives an impression that it can also be interpreted as "if," such as "if the sun rises from the east" and "if spring comes," and creates a pragmatically inappropriate meaning. Thus, the ability of *ba* to vacillate between two opposite worlds generates awkwardness in the interpretation of events in the domain of scientific facts.

An actual world situation is determinable, that is, its occurrence can be observed by the speaker due to an objectively perceptible event. Some determinable events are volitionally controllable by the speaker, such as activities, and some can be known to be true due to previous experience gained by the speaker, such as mathematical and scientific facts. Indeterminability, on the other hand, is concerned with situations where the speaker cannot know the truth or outcome of the event, that is, the speaker does not have volitional control over the event or simply does not know the truth of the event. In fact, determinability plays a significant role in defining the correct conditions for the use of the *ba* conditional. This can clearly be demonstrated in a past context since past events are realized events and are objectively perceptible. In the following section the behavior of *ba* sentences is examined on the basis of the dichotomy of determinable/indeterminable predicates.

5.3.1 *Ba* Sentences in Past Context

5.3.1.1 Indeterminability and Ba *Sentences in Past Context*

Although it has been illustrated that *ba* can be used to express general, nonhypothetical events in nonpast context, such as in sentence (55), *Haru ni nare-ba sakura ga* saku, "Cherry flowers bloom in spring," most *ba* sentences in past context create an "invited inference" that points to a different or opposite result from the one described. The following are some examples of this case:

(56) Kanojo ga iw-anake-reba daremo shinjitsu o shir-anakat-ta.

she NOM tell-NEG-COND no one truth ACC know-NEG-PST

"If she did not tell, no one would have known the truth."

(57)

Hikooki	ga	jikandoori	ni	tsuk-eba
airplane	NOM	on time	TEMP	arrive-COND
tsugi	no	hikooki	ni	maniat-ta.
next	GEN	airplane	DAT	on time-PST

"If the plane (I took) was on time, I would have been on time for the next plane."

(58)

Tenki	ga	waruke-reba,	shiai	wa	chuushi-ni-nat-ta.
weather	NOM	bad-COND	game	TOP	cancel-become-PST

"If the weather was bad, the game would have been cancelled."

Normal interpretations of these sentences are that they imply the unmentioned opposite alternative to the situation expressed in the sentence, such as: "The truth was in fact discovered by her disclosure;" "The delay of the plane caused the speaker to miss the next plane;" and "The game was cancelled due to bad weather," respectively. Sentences (56) and (57) are concerned with a single event regarding a particular person and cannot be viewed in terms of accustomed activity. In order for these sentences to receive habitual interpretations, specific context and/or grammatical devices such as the additional use of *mono-da* "used to be," could be included; otherwise, these sentences convey a counterfactual meaning. Sentence (58), on the other hand, shows a slightly different aspect in that it is possible to interpret the sentence as expressing a certain degree of habituality. This is because the bad weather pattern can be viewed as a common phenomenon that may habitually result in canceling the game. Such an interpretation causes one to envision that the speaker is remembering the past and being nostalgic about the fact that the game was cancelled every time the weather went bad. In either speaking or writing, without a device such as *mono-da naa . . .* "used to be," the sentence is not effective in expressing such feelings toward the past event.

When paying careful attention to the type of predicates used in the antecedents, one recognizes that these conditions are illustrated by stative predicates (sentences [56] and [58]) and unaccusative verbs (sentence [57]) that are considered indeterminable predicates as defined in this book.[18] These predicates probably cannot be viewed as expressing a habitual concept and play a significant role in the assigning of hypothetical meaning to *ba* sentences.[19]

5.3.1.2 Determinability and Ba *Sentences in Past Context*

Although most *ba* sentences in past context can be said to express a counterfactual event, there are some cases where *ba* can express past

18. Indeterminable, volitionally uncontrollable events such as the weather can be observed objectively as habitual phenomena. However, the point here is that without any additional information, such interpretation may be possible. For example, the accompanying *kanarazu* "without fail" *with Ame ga fur-eba koozui -ni -nat-ta* can connote the event's habitual occurrence "Every time it rained, (the river) flooded without fail," but without such an adverbial phrase, the sentence receives a counterfactual interpretation.
19. Another phenomenon that can be noticed through observation of example sentences (56)–(58) is the compatibility of the predicate in the consequent with the antecedent. When volitionally uncontrollable predicates are used in the antecedent, the predicate in the consequent is compatible with the same type of predicate. For example, *mani at-ta*, "was on time" and *shir-anakat-ta*, "did not know," in sentences (56) and (57) are negative (stative) and intransitive predicates, respectively, that are in the volitionally uncontrollable category; therefore, the sentences do not create awkwardness.

habitual events. Generality, connoted in past habitual events, often describes a speaker's emotional reminiscences. Such use of *ba* to express the emotional demeanor of the speaker is comparable to the use of *ba* in nonpast context where the speaker expects that the whole statement is valid. The spatiotemporal distance between the time of speech and the time referenced can be described as different directions relative to a boundary formed by the time of speech: the hopes/wishes/expectations of the speaker are referenced from the time of speech toward the future in nonpast context; on the other hand, the reminiscence/nostalgia of the speaker is referenced toward the past. Thus, the function of *ba* can commonly be seen in both contexts.

Determinable predicates communicate such nonhypothetical notions since they describe events that are volitionally controllable, generalizable, and/or predictable to the speaker, that is, the realization of an event can be determined by the speaker due to the knowledge the speaker possesses. The difference in the way determinability is viewed in *ba* sentences from that in which *to* and *tara* sentences are viewed is that determinability communicates habituality, but not semelfactivity, since *ba* sentences cannot be used to express a one-time event. With such circumscription, when determinable predicates are used in a *ba* construction, the *ba* sentence in past context can convey either a habitual or a counterfactual event. Such interpretation is possible given the fact that the meaning of Japanese conditional sentences is not merely dependent upon the antecedent, but the consequent also has impact on the interpretation of the sentence. The following examples prove this observation:

(59) a. Ame no hi ni ik-eba, mise wa suite-i-ta.

rain GEN day TEMP go-COND store TOP not-crowded-PERF-PST

"Every time we went to the store on a rainy day, it was uncrowded./If I had gone to the store on a rainy day, the store would not have been crowded."

b. Ototoi no ame no hi ni ik-eba,
the day before yesterday GEN rain GEN day TEMP go-COND

mise wa suite-i-ta . . .
store TOP uncrowded-PERF-PST

"If I had gone to the store the day before yesterday, which was a rainy day, the store would not have been crowded."

c. Ame no hi ni itte-ire-ba,
rain GEN day TEMP go-PERF-COND

mise wa suite-i-ta.
Store TOP not-crowded-PERF-PST

"The store would have been less crowded if I had gone on a rainy day."

(60) a. Kuruma ni nor-eba kibun ga waruku-nat-ta.
car LOC ride-COND feeling NOM bad-become-PST.

"Whenever I rode in a car, I got car sick."

? b. Anotoki kuruma ni nor-eba kibun ga waruku-nat-ta.
that time car LOC ride-COND feeling NOM bad-become-PST

"If I had ridden in a car, I would have gotten car sick."

c. Kuruma ni notte-ire-ba kibun ga waruku-nate-i-ta.
car LOC ride-PERF-COND feeling NOM bad-become-PERF-PST.

"If I had ridden in a car, I would have gotten car sick."

(61) a. Ondo o 32 do ni agere-ba, koori wa toke-ta.

temperature ACC 32 degree to raise-COND ice TOP melt-PST

"Raising the temperature to 32 degrees melted the ice./If I had raised the temperature to 32 degrees, the ice would have melted."

b. Sono jiten de ondo o 32 do ni agere-ba,

that point TEMP temperature ACC 32 degree to raise-COND

koori wa toke-ta.

ice TOP melt

"If I had raised the temperature to 32 degrees at that point, the ice would have melted."

c. Ondo o 32 do ni agete-ire-ba,

temperature ACC 32 degree to raise-PERF-COND

koori wa toke-ta.

Ice TOP melt-PST

"Raising the temperature to 32 degrees melted the ice./If I had raised the temperature to 32 degrees, the ice would have melted."

Sentences (59)a, (60)a, and (61)a are somewhat semantically ambiguous: they can be interpreted as expressing either a past habitual event or a counterfactual event. The interpretation of these sentences, however, depends on the degree of habituality perceived in the event in the antecedent. When the event in the antecedent is recognized as routine activity that anyone can perform iteratively, the *ba* sentence tends to receive a habitual interpretation. This observation is especially exemplified by sentences (59)a and (60)a whose interpretation is prone to be the habitual one, while sentence (61)a may not be recognized as habitual at a glance unless one is routinely under circumstances where ice is melted, for example, in one's work at a lab. Thus, the fact that there is a degree to which hypotheticality is detected in *ba* sentences should be acknowledged. The reason that sentence (60)b is considered awkward is because the whole event described by the *ba* sentence is best viewed as the speaker's habit rather than a one-time hypothetical event.

The habitual interpretation of these sentences can also be supported by the fact that they can accompany *itsumo*, "always." For example, sentence (61)a as a whole is concerned with a scientific fact that is proven to be true, and such relation of cause and effect should be seen equally and uniformly not only in the present/future time framework, but also in the past time framework. However, when specific context is provided, the sentences no longer express habituality, but express only counterfactuality. This is exemplified by sentences (59)b and (61)b which allow a counterfactual interpretation only. Sentence (60)b is awkward since the interpretation of the sentence leans toward the habitual one and is not reconcilable with the specificity expressed by *anotoki* "that time." This ineffectiveness can be remedied through the use of *te-iru* as seen in sentence (60)c, which is another, more effective device that allows a *ba* sentence to express a counterfactual meaning. The use of the aspectual form *te-iru* in either the antecedent or the consequent or both for nonstative predicates is a powerful method to remove the ambiguity of the sentence, as exemplified in sentences (59)c and (61)c.

The type of predicates are thus crucial to the investigation of the essential functions of *ba* sentences. It has to be mentioned, however, that there are cases where only habitual interpretation is possible, regardless of the type of predicate in the *ba* construction, for example, where the antecedent is describing a scientific fact or something that is perceived as occurring inevitably and habitually; therefore, an "if" interpretation is not applicable. In such cases, the whole antecedent is viewed in terms of determinability, as in:

(62) Yoru ga kure-ba, kossorito uchi o nukedashi-te,
night NOM come-COND secretly house ACC sneak away-GER

Momoko wa otoko no tokoro e it-ta.
Momoko TOP man GEN place LOC go-PST

"Momoko sneaked away from her house and visited her boyfriend every night."

(63) a. Haru-ni nar-eba tsubame ga ki-te,
spring become-COND swallow NOM come-GER
su o tsukut-ta.
nest ACC make-PST
"Swallows used to come and make nests in the spring."

* b. Kyonen haru-ni-nar-eba tsubame ga ki-te,
last years spring-become-COND swallow NOM come-GER
su o tsukut-ta.
nest ACC make-PST
"Swallows used to come and make nests in the spring."

(64) Fur-eba doshaburi-dat-ta.
fall-COND pouring-COP-PST
"Whenever it rained, it came down in buckets."

Sentences (62) to (64) receive a past habitual interpretation only. In regard to sentences (62) and (63), this is simply due to the common understanding that night comes everyday and spring comes every year. Even a *ba* sentence can describe some kind of scientific fact; hence, it is possible to interpret the sentence expressing a hypothetical meaning if an event in the antecedent is not necessarily realized. This instance was already seen in sentence (61)a where the speaker talks about the possibility of melting ice upon raising the temperature to 32 degrees. Although the whole sentence is concerned with a scientific fact, the event in the antecedent is still under the speaker's control and cannot necessarily be viewed from a habitual perspective. In sentence (62), however, a hypothetical situation interpreted as "If night had come, she would have gone to her boyfriend's place" cannot be envisioned, since the invited inference that "Night did not come and she did not visit her boyfriend" is hardly cognizable. The same can be observed for sentence (63)a, since scientific natural phenomena cannot express counterfactuality, for example, when accompanying the temporal element *kyonen* "last year," sentence (63) is simply inappropriate in that the antecedent describes the arrival of spring as a routine phenomenon and is incompatible with a specific occasion.

Although native speakers of Japanese can naturally comprehend the habitual interpretation of sentence (64), it is quite challenging to explain what factor causes the sentence to receive such an interpretation, since the antecedent is not something that can be perceived as habitual, as in sentences (62) and (63)a. The explanation lies in the lack of cause and effect. "Rain" and "pelting heavy rain" are identical entities that are occurring simultaneously; thus, the consequent is describing the antecedent in that the heavy volume of rain itself is the phenomenon of rain. For example, it is easy to perceive that a flood is caused by a large volume of rain, and that a stomach ache is caused by the result of eating uncooked chicken, and so forth. If the sentence describes some sort of contingency relationship between the antecedent and the consequent, then the hypothetical interpretation—that is, "If it had~, then it would have been~,"—would have been easily comprehended.

Thus, in order for a *ba* sentence to receive only a nonhypothetical interpretation—and a hypothetical interpretation is in no way possible—the event in the antecedent has to be concerned with a phenomenon that is inconceivable not to envision as habitual—that is, events that inevitably occur on a regular basis together with the notion of determinability are key to such an interpretation.

The notion of determinability is thus crucial to determining the meaning of *ba* sentences in past context: indeterminable predicates have an influence over the hypothetical interpretation, while determinable predicates can render *ba* sentences to express either a counterfactual meaning or a habitual meaning. Which interpretation prevails over the other depends on the way we perceive the whole sentence: if a sentence is perceived as a routine event, the habitual interpretation prevails; if a sentence is perceived as specific event, the sentence expresses hypotheticality. Here again, when *ba* sentences are viewed in terms of generality, one of the functions of *ba* and *to* is identical in that both can be used to express habitual, general events in past contexts. For example, *Haru ni nare-ba kanarazu tsubame ga ki-ta* and *Haru ni naru-to kanarazu tsubame ga ki-ta* translate into the same English sentence, "Whenever spring came, swallows came without fail." The difference is that *ba* sentences relate a speaker's nostalgia toward a past event that no longer occurs in the present time, whereas *to* sentences do not relate such an implication. This idea of "no longer the case" generated by *ba* sentences also communicates counterfactuality at root and establishes an invited inference pragmatically revealing to us that, in reality, the event resulted in a different way than mentioned in the sentence itself.

5.3.2 *Ba* Sentences in Nonpast Context

Since events in nonpast context are largely indeterminable and not objectively perceptible, one might expect a degree of difficulty in detecting nonhypotheticality and determining the behavior of *ba* sentences in nonpast context. The following section examines the issue of indeterminability/determinability in relation to such.

5.3.2.1 Indeterminability and Ba *Sentences in Nonpast Context*

It was demonstrated that, in a past context, predicates that represent the notion of indeterminability (cannot be known/predicted, cannot be volitionally controlled) mostly render *ba* sentences to receive a hypothetical interpretation. The behavior of *ba* may be found to be the same as that used in a past context, and this assumption can easily be justified since the notions of indeterminability and nonpast context altogether are concerned with nothing but an unknown, unrealized situation. The following examples illustrate this observation:

(65)	Taifuu	ga	kure-ba,	ensoku	wa	chuushi-da.
	typhoon	NOM	come-COND	field trip	TOP	cancel-COP

"If the typhoon comes tomorrow, the field trip will be cancelled."

(66)	Shigoto	ga	owar-eba	eiga ni	ik-eru.
	job	NOM	finish-COND	movie DAT	go-POT

"If the job is done, I can go out to a movie."

(67)	Kare	ga	nihongo	de	hanas-eba	wakaru.
	he	NOM	Japanese	INS	speak-COND	understand

"If he speaks in Japanese, I will understand."

(68)	Choojoo	ni	tsuk-eba	gekai	ga	mioros-eru.
	summit	LOC	arrive-COND	the earth	NOM	look down-POT

"If you reach the summit, you can have a bird's-eye view of the earth."

Sentences (65)–(68) all express hypotheticality, and it is difficult to perceive of these sentences as expressing a habitual meaning. This difficulty is verified by the fact that these sentences are awkward when accompanying the adverb *itsumo* "always" that expresses a high frequency of the event's occurrence. A native speaker of Japanese would automatically employs the *to* conditional to express a habitual meaning. In sentence (65), for example, the unknown situation is whether or not the typhoon comes at the time of the field trip, which cannot be perceived as a generalizable event. This observation can also be applied to sentences (66) to (68), none of which indicate a habitual event; rather, they imply the opposite possible alternative—an invited inference—and in order to receive a general interpretation, grammatical devices such as *atarimae-da* and *futsuu-da* "it is naturally the case that ~" are required. Thus, when a predicate is concerned with indeterminability in which a speaker cannot recognize an objectively perceptible event, the *ba* sentence receives a hypothetical interpretation. This result allows one to easily predict the behavior of indeterminable predicates in *ba* sentences in that they behave in the same fashion as those in past context: the whole sentence can refer to either a hypothetical world situation or a nonhypothetical world situation depending on context.

5.3.2.2 Determinability and Ba *Sentences in Nonpast Context*

Even among native speakers of Japanese the fact that *ba* sentences can be used to express nonhypothetical situations (generally understood events) is often overlooked. Nevertheless, one notes that *ba* sentences in nonpast context can also express a general event that is commonly understood as a fact or as an individual's habitual activity, just as *ba* sentences can express such notions in past context. In the following examples, determinable events are examined in *ba* constructions to exhibit the influence of such events:

(69) Karushium o takusan sesshu sure-ba hone ga joobu-ni-naru.
calcium ACC plenty take-COND bone NOM strong-become
"If I/you take lots of calcium, my/your bones get stronger./Calcium makes your bones strong."

(70) Hoomu sutei o sure-ba kaiwa ga jootatsu suru.

homestay ACC do-COND conversation skill NOM improve

"Staying with a family will improve one's conversation skills./If I/you stay with a family, my/your conversation skills will improve."

(71) Inu wa mikka ka-eba sannen on o wasure-nai.

dog TOP three days have-COND three years indebt ACC forget-NEG

"Dogs do not forget a debt of gratitude of their owner./If I/you take care of a dog for three days, he will not forget a debt of gratitude for three years."

As the English equivalents show, sentences (69)–(71) can be considered as presenting a speaker's general knowledge that the consequent will be realized upon the realization of the antecedent. In other words, these sentences can also be used to refer to known facts, observable situations, that is, those in the domain of actual worlds. For example, in sentence (69), it can be acknowledged that the speaker is sharing the information s/he possesses about calcium with the interlocutor. Under such interpretation the degree of hypotheticality expressed in the antecedent is not a significant issue. Evidence can be found in the rather weak interpretation of the invited inference: "If you do not take calcium, your bones will not become strong." Unlike the counterfactual interpretation of the sentence "If I had money, I would have bought a new car," in which a speaker is most likely trying to convey the fact that "I did not buy a new car, since I did not have money," the speaker in sentence (69) is not trying to impel the interlocutor to envision a different situation by uttering the sentence. Perhaps the interlocutor does not take "Your bones will not get stronger if you do not take calcium" as the real message. The same thing can be observed for sentences (70) and (71), where in the former, it is not necessarily true that a person's language does not improve unless s/he homestays, and in the latter, we are admonished not to forget what others have done for us, taking a dog as an example. In this type of common saying, even an opposite possible alternate cannot be envisioned. Sentences (69)–(71), therefore, can be said to express general knowledge possessed by the speaker.

Nevertheless, when these sentences co-occur with an element that expresses specificity, they no longer convey general knowledge, but express hypotheticality. The following are examples of such cases:

(72) Ima shikago ni ik-eba buruzu no shiai ga mi-rareru.
now Chicago LOC go-COND Bulls GEN game NOM watch-POT
"If you go to Chicago now, you can go see a Bulls game."

(73) Kono hon-ja-naku-te sono hon de shirabere-ba
this book-COP-NEG-GER that book INS search-COND
kotae ga wakaru.
answer NOM find out
"If you search using that book, not this book, you will find out the answer."

Here, specific elements remove the nonhypotheticality and cannot communicate generality. Both sentences create an invited inference that "You will miss the Bulls game unless you leave for Chicago now" and "You may not find the answer if you use this book," respectively. Such *ba* sentences receive a hypothetical interpretation only.

The feature of the antecedent is the key for such sentences to receive a hypothetical interpretation. Here again, nonhypotheticality expressed by *ba* sentences can be viewed as similar to that of *ba* sentences in past context. When the antecedent is concerned with some sort of natural phenomena that inevitably realize on a regular basis, the *ba* sentence unavoidably expresses nonhypotheticality only. Such examples are:

(74) Hi ga nobor-eba atatakaku-naru.
the sun NOM rise-COND warm-become
"When the sun rises, it becomes warm."

(75) Fuyu wa 5 ji-ni-nar-eba moo kurai.
winter TOP 5 o'clock-become-COND already dark
"It gets dark around 5:00 in winter."

The notion of hypotheticality can be applied to neither sentence (74) nor (75), since the antecedent in both sentences inevitably occurs regularly and we view such phenomena as generic. Also, there is no invited inference derived from these sentences; the sunrise and the time reaching 5 o'clock is expected to occur every day.

It is therefore clear that the notions of indeterminability and determinability are crucial to the interpretation of *ba* sentences, and whether a *ba* sentence expresses a hypothetical or nonhypothetical meaning depends on the feature of the event in the antecedent. Since indeterminability at root communicates hypotheticality, it is straightforward to see that indeterminable predicates contribute to the hypothetical meaning expressed by *ba* sentences. However, most determinable events, except those represented by sentences (74) and (75), are conceived as expressing either hypothetical or nonhypothetical meanings just as those in past context. In past context, such vacillation can be explained from the viewpoint of an "unrecoverable event" that is considered a common thread to both "whenever" and "if" situations.[20] In nonpast context, however, the notions of hypotheticality and nonhypotheticality cannot be projected from such a viewpoint, since the future is after all unexplored.

The common thread of hypothetical and nonhypothetical interpretations that determinable events create may be traced in a weak interpretation of invited inference. Clearly, the nonhypothetical interpretation, "it is naturally the case that ~," does not generate an invited inference, as exemplified in sentences (69)–(71). Furthermore, a hypothetical interpretation of these sentences does not make one envision an invited inference: the implication is not what the speaker is actually trying to convey. This weak degree of invited inference seems to originate from the basic signification of "determinability" that is

20. "If I had money, I would have bought a car," in a pragmatic view, implies that the speaker did not buy a new car in reality at the time of speech. Also, "Whenever I had money, I gave it away/I used to give money away whenever I had it" implies, in a pragmatic view, that the speaker no longer gives money away. The common notion of these sentences is that the situation at the time of speech is, in fact, an unmentioned opposite or different situation from what the sentences actually state.

expressed by speaker confidence/knowledge in the realization of the event described by a *ba* sentence.

The notions of determinability and indeterminability are also the key to approaching a unique feature of *ba* sentences—the fact that some *ba* sentences do not allow the co-occurrence of a volitional expression in the consequent—a problem that has previously not fully resolved. In the following sections, the focus of the investigation shifts to this problem, concerned with the compatibility of the antecedent and the consequent in *ba* sentences.

5.3.3 *Ba* Sentences, Modality Expressions, and Negation

5.3.3.1 Ba *Sentences and Volitional Expressions*

Inaba (1991) claimed that *ba* following a nonstative predicate does not allow the co-occurrence of modality expressions in the consequent, while *ba* following a stative predicate does not have such as constraint in the consequent clause. For example, *Shikago e ik-eba buruzu no shiai o mi-yoo* "If I go to Chicago, I will see a Bulls game" is incorrect since *ba* accompanies the nonstative predicate *iku* "to go." On the other hand, *Samuke-reba mado o shimete-kudasai* "If it is cold, please close the window" is grammatically correct. Inaba approached this from the viewpoint of divergent phenomena: one particular form in the learner's native language corresponds to more than two forms in the learner's target language. She pointed out the grammatical difference between Japanese conditionals and English conditionals.

The volitional expressions in question here are commonly accepted as those that express notions such as will, wish, request, command, and so forth. The equivalents of such expressions in Japanese are, for example, *kaku,* "I will write," *kak-oo,* "I shall write," *kaki-tai,* "I want to write," *kaite-kudasai,* "please write," *kak-e,* "write," and so forth. These volitional expressions are merely used to express a speaker's state of mind; they are not-yet-realized and therefore express ungeneralizable situations. This notion of "ungeneralizability" communicates the idea of hypotheticality in that both exclusively exist in the domain of nonactual world situations.[21]

The above observation leads us to the hypothesis that volitional expressions may be compatible with a *ba* construction when the

21. In fact, it should be noticed that most volitional expressions do not form a past tense form.

antecedent fulfills the condition that the whole *ba* sentence expresses a nonactual world situation. In other words, when indeterminable predicates—which render a hypothetical interpretation—co-occur with the *ba* conditional, the antecedent can invite a volitional expression in the consequent. On the contrary, when a determinable predicate—which renders a nonhypothetical interpretation—co-occurs with the *ba* conditional, the antecedent cannot invite a volitional expression in the consequent.

The current view of the stative and nonstative predicate dichotomy is not, however, perfect justification in distinguishing the behavior of volitional expressions. There are nonstative predicates used in *ba* constructions that can invite volitional expressions in the consequent, as shown in the following examples:

(76)	Kono shigoto	ga	owar-eba	sono shigoto	o	suru	tsumori-da.
	this job	NOM	finish-COND	that job	ACC	do	intend-COP

"If this job is finished I will do that job."

(77)	90 ten	ijoo	tor eba	A	o	age-mashoo.
	90 point	over	get-COND	A	ACC	give-VOL

"If you score over 90 points, I shall give you a grade of A."

(78)	Anata	ga	nom-eba	watashi	mo	nomu.
	you	NOM	drink-COND	I	also	drink

"If you drink it, I will drink too."

The common point of the predicates used in these antecedents is the lack of determinability over the situation described in the antecedents. For example, the predicate *owaru*, "to be finished," in sentence (76) is an unaccusative verb whose outcome cannot be known to the speaker. The same thing can be observed for sentences (77) and (78). Both *90 ten toru*, "score 90 points," and *anata ga nomu*, "you drink," represent some other individual's activity and are beyond the speaker's control. All of these antecedents thus include indeterminable predicates, which, in turn, communicate hypothetical, nonactual world situations, and demonstrate that these predicates in *ba* constructions can invite a volitional expression in the consequent. Contrary to this, when a determinable predicate is used in the antecedent, it can no longer invite a volitional expression in the consequent. The following are examples of this case, where both sentences (79) and (80) include an activity predicate in the antecedent and a volitional expression in the consequent. Neither makes sense:

(79) * Shikago e ik-eba buruzu no geemu o mi-yoo
Chicago LOC go-COND Bulls GEN game ACC see-VOL
"If I go to Chicago, I will see a Bulls game."

(80) * 200 man en tamer-eba kuruma ga kai-tai
two million yen save-COND car NOM buy-DES
"If I save two million yen, I want to buy a new car."

As discussed in the previous section, determinable predicates can render *ba* sentences to express either a hypothetical or a nonhypothetical meaning depending on context and the consequent, which means that the equivalent of *ba* sentences is not necessarily always "if." This is especially exemplified in the examination of *ba* sentences in past context. Such behavior of *ba* also agrees in nonpast context. For example, *Ano mise e ik-eba Yoko san ni a-eru,* "If you go to that restaurant, you can see Yoko/People can see Yoko in that restaurant," can be interpreted as expressing the hypotheticality of someone's going to that restaurant or the general activity of an "anyone" subject, depending on context. This premise is not the case for sentences (79) and (80), neither of which can accept volitional expressions. This is due to the fact that the speaker's own activity can be determined by his/her volition, and that contradicts hypotheticality. Also, when a sentence is considered as having an "anyone" subject, expressing generality, it does not allow a volitional expression in the consequent since an "anyone" subject communicates the notion of generality, and, therefore, it is not in accord with indeterminability.

Thus, the co-occurrence of volitional expressions in the consequent of *ba* sentences is not merely due to the stative versus nonstative dichotomy; rather, it is due to the existence/nonexistence of the determinability detected in the event in the antecedent. Stative predicates, such as adjectives and existential verbs, are certainly in the category that allow invitation of a volitional expression in the consequent.

5.3.3.2 Problems with Deontic Constructions in Ba *Sentences*

Investigation of the behavior of *ba* in relation to volitional expressions shows that *ba* expresses hypotheticality in not-yet-realized situations, which themselves at root communicate volitional expressions. However, there is a problem observed in deontic constructions, especially regarding the command/request form concerned with the notion of deontic necessity. Such constructions are often extremely awkward when employed in the consequent, even when the *ba* antecedent expresses hypotheticality. Considering first the command/request forms, and then the permissive forms concerned with the notion of deontic possibility, we have the following survey examples:

(81) * a. Furui-no o suter-eba atarashii-no o kai-nasai.
old-COMP ACC discard-COND new-COMP ACC buy-IMP
"If you discard an old one, buy a new one."

b. Furui-no o suter-eba atarashii-no o katte-mo-ii.
old-COMP ACC discard-COND new-COMP ACC buy-even if-okay
"If you discard an old one, you may buy a new one."

(82) * a Honda san ni a-eba yoroshiku itte-kudasai.
Honda DAT see-COND give regard-please
"If you see Mr. Honda, please say hello to him."

? b Honda san ni a-eba kisu o shite-mo-ii.
Honda DAT see-COND kiss ACC do-even if-okay
"If you see Mr. Honda, you may kiss him."

(83) * a. Toyota san ga kure-ba kaer-i-nasai.
Ms. Toyota NOM come-COND go home-IMP
"If Ms. Toyota comes, go home."

b. Toyota san ga kure-ba kaette-mo-ii.
Ms. Toyota NOM come-COND go home-even if-okay
"If Ms. Toyota comes, you may go home."

(84) a. Okane ga ar-eba kuruma o kai-nasai.
money NOM have car ACC buy-IMP
"If you have money, buy a car."

b. Okane ga ar-eba kuruma o katte-mo-ii.
money NOM have-COND car ACC buy-even okay
"If you have money, it is okay if you buy a car."

(85) a. Samuke-reba mado o shimete-kudasai.
cold-COND window ACC shut-please
"If you are cold, please shut the window."

b. Samuke-reba mado o shimete-mo-ii.
cold-COND window ACC shut-even if-OK
"If you are cold, you may shut the window."

Although factors expressing hypotheticality exist in each of the antecedents in the above examples, sentences (81)a–(83)a are considered inappropriate. Sentences (84)a and (85)a, on the other hand, are all deemed appropriate. In fact, among all four conditionals, *tara* is the only conditional that can be used appropriately in sentences (81)a–(83)a. One significant difference between sentences (81)a–(83)a and (84)a and (85)a is that the predicates in the former group are nonstative, whereas the predicates in the latter group are stative. This demonstrates that the stative/nonstative dichotomy certainly plays a role in defining the appropriate use of the deontic forms employed in the consequent. This begs the question, however, what is the characteristic of stative predicates that makes their use with deontic forms appropriate?

The crucial difference between stative and nonstative predicates is that the latter indicate a point in time where the event in question is realized, whereas the former do not indicate a distinct/conceptual point of realization. Command/request forms are employed when a speaker demands that an interlocutor carry out a particular event—that is, the necessity of the event's realization is implied by the speaker.[22] When a nonstative predicate is used in the antecedent, and a command/request form appears in the consequent, the necessity of the event's realization described by the command/request form casts the whole sentence as a sequence of one-time events carrying a "when" meaning, since the nonstative predicate indicates a point in time where each event is realized. Furthermore, since *ba* cannot be used to express a one-time event that focuses on the sequentiality expressed by "when,"[23] a *ba* antecedent cannot co-occur with a consequent that includes a deontic form.

On the other hand, a distinct/conceptual point of realization of the event cannot be observed in relation to stative predicates. For example, in *Samuke-reba mado o shimete-kudasai,* "If you are cold, please shut the window," the sentence is not a sufficient basis for a concrete, one-time event expressed by "at the time when," due to the fact that a clear point of realization of "coldness" cannot be envisioned, even at the time the event described by the command/request form is realized. Hence, "at the point when it is cold" is a difficult concept to capture, and "if it is cold" is the more appropriate interpretation, since the more hypothetical the situation becomes, the greater the spatiotemporal distance created between the situation and the speaker, thereby granting the speaker a degree of temporal latitude in which to perceive the situation. *Ba* is therefore appropriate in such environments that express a hypothetical meaning.

Regarding the permissive form, however, *-te-mo-ii* can occur in the consequent for all sentences (81)b–(85)b. Sentence (82) is perceived as a bit awkward, likely attributable to the content of the sentence. The key here is the nature of the permissive form, which is concerned with

22. In this sense, these forms have a different feature from other volitional expressions such as *tai* "want to," *-tsumori-da* "will," and *-(y)oo* "shall" that do not necessarily involve other individuals in order to realize the event in question.

23. It is now easy to understand the reason that *tara* can be used in sentences without causing awkwardness. *Tara* places emphasis on the contingency relationship for a one-time event whereby the event in the consequent is realized upon the realization of the event in the antecedent.

possibility, not necessity. Compared to the command form, which forces one to realize the event in question, the permissive form communicates the notion of a possible/nonactual world in which the speaker merely offers a possibility of realization to the interlocutor; therefore, the decision regarding the realization of the event is left to the interlocutor, and the concept held by the permissive form in the consequent is compatible with the hypothetical meaning expressed in the antecedent.

5.3.3.3 The Ba *Construction and Negation*

It has been pointed out that *nai* and *te-iru* share a common feature in that they are homogeneous. In hypothetical interpretations, *ba* sentences do not suggest that the causal relation between the antecedent and the consequent is actual, but expresses a not-yet-realized connection between the two. The degree of hypotheticality increases when stative *te-iru* participates in *ba* sentences. This is especially true in past contexts where *te-iru* plays a vital role in distinguishing a habitual event from a counterfactual event as seen throughout the examination of *ba* conditional sentences. The static state expressed by *te-iru* is beyond a speaker's control, and the speaker cannot determine on the basis of his/her own volition whether or not the situation will change. Thus, the notions of uncontrollability and indeterminability are deeply related to hypothetical interpretations.

Nai, which can be treated as being at once both an adjective (morphologically) and a negative suffix of a predicate, was found to have semantic features similar to *te-iru* in terms of (prohibitional) energy input as discussed in chapter 4. These features of *nai* allow for the view that its behavior in conditional *ba* sentences may agree with both that of corresponding affirmative forms and perpetual states expressed in adjectives. In this section, the stative character of the predicate *nai* will be further examined in relation to *ba* conditional sentences.

By examining the negative forms of nonstative volitional predicates in *ba* constructions, the following examples shed some light on the question of the stativity of *nai* in *ba* constructions.

(86)	a.	Oosaka	e	ik-anake-reba	Tookyoo	e	ik-e-ru.
		Osaka	LOC	go-NEG-COND	Tokyo	LOC	go-POT

"If I (you) do not go to Osaka, I (you) can go to Tokyo (with that money)."

*	b.	Oosaka	e	ik-anake-reba	Tookyoo	e	ik-oo.
		Osaka	LOC	go-NEG-COND	Tokyo	LOC	go-VOL

"If I do not go to Osaka, I will go to Tokyo (instead)."

*	c.	Oosaka	e	ik-anake-reba	shigoto	o	shite-kudasai.
		Osaka	LOC	go-NEG-COND	work	ACC	do-please

"If you do not go to Osaka, please work."

(87)	a.	Himitsu	o	baras-anake-reba	inochi	wa	tasukaru.
		secret	ACC	tell-NEG-COND	life	CNT	saved

"If I (you) do not tell the secret, my (your) life will be safe."

* b.	Shinjitsu	o	iw-anake-reba	wakare-yoo.
	truth	ACC	tell-NEG-COND	separate-want

"If I do not tell (you) the truth, let's break up."

* c.	Shinjitsu	o	iw-anake-reba	keemusho	ni	i-nasai.
	secret	ACC	tell-NEG-COND	jail	LOC	stay-IMP

"If you do not tell us the truth, stay in jail."

Sentences (86)a, b, and c, and (87)a, b, and c respectively involve potential, desire, and command/request expressions in their consequents. It can easily be seen that the behavior of *nai* is identical to its corresponding affirmative forms, indicating that negative forms of nonstative volitional predicates can be controlled and determined by the speaker in the same manner as in the corresponding affirmative forms. Both "going" and "not going" can volitionally be controlled and determined by the speaker. This is one reason why *nai* behaves less like a stative predicate and more like a nonstative predicate as seen in (86) and (87). This controllability and determinability excludes hypotheticality from associated *ba* sentences, which do not allow the co-occurrence of a negative antecedent with the desiderative expression or the command/request expressions. This feature of *nai* is the same as that of corresponding affirmative forms and is further illustrated by observing the following examples:

(88) a. Kono shigoto ga owar-eba tsugi no o shi-yoo.
this job NOM finish-COND next one ACC do-VOL
"If this job is not finished, I will not do that job."

b. Kono shigoto ga owar-anake-reba tsugi no wa shi-nai.
this job NOM finish-NEG-COND next one CNT do -NEG
"If this job is not finished, I will not do that job."

The nonstative verb *owaru* "to finish" (unaccusative verb) was one of the predicates that justified the claim that the co-occurrence of volitional expressions in the consequent of *ba* sentences is due to the existence/nonexistence of the determinability detected in the event in the antecedent. The negative form of *owaru* in sentence (88)b is considered perfectly appropriate.

It is clear that the corresponding affirmative predicate determines the behavior of *nai* predicates in *ba* constructions. If the affirmative form of a predicate cannot be volitionally controlled and is indeterminable by the speaker, a *ba* construction can invite a volitional expression in the consequent. From this observation, it is easy to predict that the negative form of a stative predicate can invite any type of expression in a *ba* consequent. The following examples confirm this:

(89) a. Oishiku-nake-reba takusan nokoru.
delicious-NEG-COND a lot remain
"If it's not delicious, I (you) will have a lot of leftovers."

b. Oishiku-nake-reba tabe-nai.
delicious-NEG-COND eat-NEG
"If it's not delicious, I will not eat."

c. Oishiku-nake-reba sutete-kudasai.
delicious-NEG-COND dispose-please.
"If it's not delicious, please dispose of it."

(90) a.	Tabemono	ga	nake-reba	iki-rare-nai.
	food	NOM	exist-NEG-COND	live-POT-NEG
	"If we do not have food, I (you) cannot live."			
b.	Tabemono	ga	nake-reba	ik-anai.
	food	NOM	exist-NEG-COND	go-NEG
	"If there is no food, I will not go (to the party)."			
c.	Tabemono	ga	nake-reba	katte-kite-kudasai.
	food	NOM	exist-NEG-COND	buy-come-please
	"If there is no food, please go and get some."			

The negative form *nai*, due to its morphology, might be expected to behave in the same manner as stative predicates, which are volitionally uncontrollable. Hence, negative forms in *ba* constructions might be expected to be capable of inviting volitional expressions in their consequents. It was found, however, that the corresponding affirmative form determines the behavior of *nai* when used in a subordinate clause concerned with modality (conditional *ba* constructions). In hypothetical sentences, the controllability and determinability of a predicate in the antecedent is not dependent on its form, but on its meaning. Therefore, even though *nai* itself is a stative predicate, this stativity does not affect the degree of hypotheticality in *ba* sentences; rather, it is affected by the type of the corresponding affirmative. *Nai* attached to these predicates plays the same role as an affirmative form, most noticeably in terms of the speaker's volitional control.

5.3.3.4 Summary of Ba *Sentences, Modality Expressions, and Negation*

It is quite clear that the compatibility of volitional expressions with *ba* constructions is dominated by the nature of the predicate and its environment in the antecedent. This characteristic is not merely dependent on the dichotomy of stativity and nonstativity; rather, it relies strongly on the speaker's volitional control over the realization of events and the necessity or possibility of the event's realization as viewed by the speaker. A *ba* construction including a predicate that is volitionally controllable by the speaker tends to receive a general, habitual interpretation unless a specified element is emphasized. This was seen in the example *Karushiumu o takusan sesshu sure-ba hone ga joobu-ni-naru* "(It is naturally the case that) bones get stronger by taking a lot of calcium." In this case, the verb *joobu-ni-naru* "become stronger" in the consequent is understood as expressing a nonspecific, general event that is compatible with such an antecedent. The sentence strongly communicates the notion of epistemic "necessity" in that the proposition is necessarily true from the speaker's viewpoint. Nevertheless, volitional expressions that refer to not-yet-realized situations exist in possible/nonactual worlds and, therefore, are incompatible with an antecedent in a volitionally controllable predicate. This idea, however, may be reversed when the antecedent involves an element that is beyond the speaker's control, such as another individuals' activities and natural phenomena. For example, once *anata* "you" is added to the antecedent *karushiumu o takusan sesshu sure-ba* "take a lot of calcium," the event is no longer in the speaker's control, and it is open to a hypothetical interpretation. The hypothetical meaning is compatible with the not-yet-realized situation in that both refer to nonactual worlds. This is why a *ba* construction with a volitionally uncontrollable predicate can invite a volitional expression, such as *Anata ga ik-eba watashi mo iku* "If you go, I will go too."

The volitional controllability of human beings also becomes a basis for why command/request forms, which express the notion of deontic necessity, behave in a different way from other volitional expressions. In volitionally controllable events it is easier to comprehend a distinct point in time for the event's realization. Command forms appearing in the consequent of a *ba* sentence entice the event in the antecedent to actualize as a sequential event expressing a "when" meaning. The notion of necessity implied by "when," however, cannot be perceived as expressing a general event, since command forms cannot be used to indicate such. A *ba* antecedent, therefore, cannot be used with a command form in the consequent. This reasoning does not apply to stative predicates in the antecedent, because a point of realization of an event cannot be recognized in stative predicates, and therefore, sequentiality is less distinct when *ba* appears in the antecedent, as seen in *Samuke-reba mado o shimete-kudasai,* "If it is cold, please shut the window."

The negative form *nai,* in fact, structurally vacillates between two domains, as observed in internal and external negation structures. Whether or not this leads to nonhypotheticality or hypotheticality, the latter expressing modality, depends on features of the corresponding affirmative form. The impracticality of negative *nai* in expressing modality is thus seen in its inability to express a higher degree of hypotheticality in *ba* conditional sentences. Unlike stative *te-iru, nai* itself does not have a significant effect on whether a *ba* sentence receives a hypothetical interpretation. *Nai* attached to a predicate is, after all, a suffix that does not function independently, but one that is determined by the corresponding affirmative form.

Stativity thus has a strong influence over the determination of hypothetical readings in that it plays a role in increasing the degree of hypotheticality and, therefore, the degree of modality. Stativity expressed by aspectual categories, however, plays such a significant role only when expressing a pure static state where no change, no dynamics, and no energy input is involved in a situation. The degree of modality becomes high upon reference to these static worlds.

Thus, human volition forms a considerable portion of the solution to the question of what makes one perceive hypotheticality or nonhypotheticality. Volitional controllability communicates the way humans operate upon events—it can be changed, or it can remain constant. Human operations, after all, have an effect on the real world (actual world situations), and as such cause *ba* conditional sentences to receive a nonhypothetical interpretation. Elements beyond our control, however, are exactly that; hence, there exists a spatiotemporal distance between us and those situations. One can merely speculate on the consequences of events existing in nonactual world situations, which relates directly to the ability of *ba* to express hypotheticality.

5.3.4 Summary of the *Ba* Conditional

The examination of *ba* demonstrates that *ba* sentences emphasize neither a natural course of events (as *to* sentences do), nor a one-time event (as *tara* sentences do). *Ba* sentences express either hypothetical events or general events in both nonpast and past contexts. Whether a *ba* sentence receives a nonhypothetical or hypothetical interpretation is predominately controlled by the features of the events in the antecedent.

Determinable events cause *ba* sentences to convey either a previously actualized situation (nonhypothetical interpretation) or a not-yet-realized situation (hypothetical interpretation). Which interpretation prevails depends on the way that the whole sentence is perceived: if a sentence is perceived as a routine event, then the habitual interpretation prevails and such interpretation is concerned with the notion of epistemic necessity; if a sentence is perceived as a specific event, then the hypothetical interpretation prevails, concerned with the notion of epistemic possibility.

On the other hand, indeterminable events such as unaccusative events, natural phenomena, and another individual's activity, which are perceived as volitionally uncontrollable by the speaker, naturally cause *ba* sentences to convey a nonactual world situation (i.e., a hypothetical interpretation). Although specificity does not grant a *ba* sentence the ability to convey a one-time event in the actual world situation in the way *tara* sentences can, it instead communicates hypotheticality in a *ba* sentence.

Such features of predicates are the basis not only for the interpretations of *ba* sentences, but also for determining the appropriate/inappropriate use of some deontic forms in the consequent. When a predicate used in the antecedent is nonstative, deontic forms take advantage of the predicate's ability to force the whole sentence to be interpreted as one specific event that necessarily will be realized. However, this is not the case for permissive forms that are concerned with the notion of possibility. Permissive forms are reconcilable with hypotheticality since *ba* can express hypotheticality regardless of the type of predicate used in the antecedent.

It should be mentioned that the indeterminability captured in *ba* sentences does not wholly overlap with the way it is captured in *tara* sentences. In the latter, volitional uncontrollability is intertwined with semelfactivity and is crucial to determining the meaning of *tara* sentences. However, such is not an essential factor for *ba* sentences due to the ability of *ba* to express generality in both past and nonpast contexts, which is not expressed by *tara*. Even if an event is volitionally uncontrollable, unlike in nonpast context, past context allows the speaker to view such an event externally and objectively, that is, the speaker can recognize the event's occurring iteratively. Since modality is concerned with the speaker's belief and knowledge, the use of conditionals cannot be viewed from the interlocutor's viewpoint, and the use of conditionals always reflects the speaker's knowledge.

5.4 SUMMARY OF MODALITY AND CONDITIONALS

The examination of three Japanese conditionals demonstrated that each has a bipolar function, expressing either a hypothetical or nonhypothetical meaning depending on tense and the type of event in the antecedent. Hypotheticality originates from the notion of indeterminability, where the realization of the event cannot be known, predicted, or controlled by a speaker. Nonhypotheticality, on the other hand, originates from the notion of determinability, where the realization of the event can be known, predicted, or volitionally controlled by the speaker. This notion of determinability is equivalent to a "when" interpretation that is concerned with a one-time event, and a "whenever/every time" interpretation that is concerned with a habitual event. Through the examination of three conditional forms it was found that they have an ability to express two of these meanings, but those two meanings do not overlap with each other.

To can express either a hypothetical or a nonhypothetical meaning in both past and nonpast contexts. In this examination, *to* is viewed as predominantly expressing nonhypotheticality. However, as is seen in examples (15), (16), (22) and (23), there are certainly cases where *to* is used to describe hypothetical situations. Even though *to* sentences convey the speaker's confidence in the realization/truth of the event, unpredictable, ungeneralizable events cannot defeat such strong features of *to* sentences. This is clearly seen in the use of *to* sentences in past context. Unless grammatical devices or phrases, such *as yokat-ta,* are provided, a hypothetical interpretation is rarely envisioned. *To* sentences in past context therefore can be said to lose the secondary ability to express hypotheticality in historicalized worlds. It is clear that *to* operates primarily upon facts and, compared to the other two conditionals, is most remote from the notion of modality. This is also in part due to the fact that *to* covers all three meanings, covering various language phenomena objectively.

Tara can express either a one-time or hypothetical meaning in nonpast context; hence, when the speaker cannot predict or control the event's realization in the antecedent, the *tara* sentence receives a hypothetical interpretation. On the other hand, if prediction or volitional control over the situation is possible, the sentence most likely receives a one-time event interpretation.

Ba, however, cannot express a one-time event; instead, it points to a hypothetical interpretation in both past and nonpast contexts. When the event is determinable from the speaker's viewpoint, *ba* sentences express a habitual meaning.

The relationship created by these two notions (hypothetical and nonhypotheticality) among the three conditionals is represented in the following illustration:

(91)

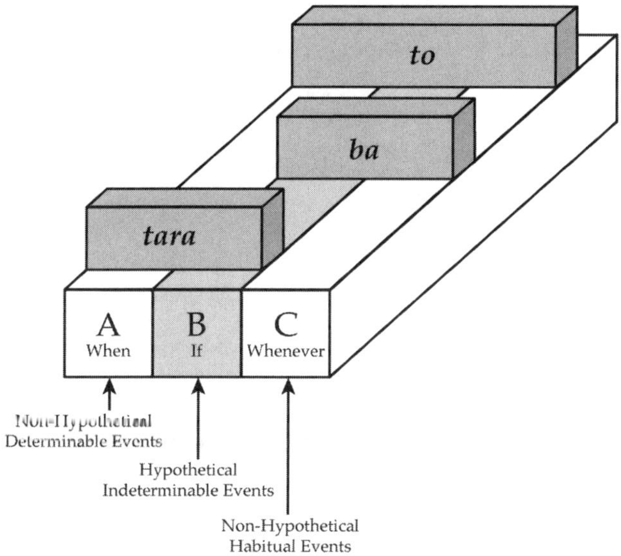

The domains, "A," "B," and "C," (representing determinable, indeterminable, and habitual events, respectively) are, of course, not clear-cut and do not reflect a precise ratio of the meanings expressed by these conditionals; however, the chart does provide an illustration for purposes of visual representation. In considering this diagrammatic representation:

Tara spans the domains of A and B:

A. Past or nonpast events expressing a "when" meaning. Nonhypothetical future interpretation is due to the speaker's self knowledge.

Uchi kaet-tara denwa shimasu.

"When I get home, I will call you."

Hontoo no koto o it-tara naguraret-a.

"When I told the truth, I was beaten up."

B. Nonpast, hypothetical events expressing an "if" meaning due to the indeterminable nature of the events.

Dame da to iware-tara akirame-yoo.

"If I am rejected, I will give up."

Ame ga fut-tara, shiai wa chuushi-da.

"If it rains, the game will be cancelled."

Ba spans the domains of B and C.

B. Past or nonpast hypothetical events expressing an "if" meaning. When the event is in past context, it expresses a counterfactual meaning.

Kono mondai ga toker-eba, shiken wa kanpeki-da.

"If I can solve the problem, I will get a perfect score."

Sonna koto iw-anake-reba, koto ga umaku hakon-da daroo ni . . .

"If you did not say things like that, things would have gone smoothly."

C. Past or nonpast habitual events expressing a "whenever/every time" meaning. The habituality makes *ba* sentences open to a general interpretation.

Tonari no ko wa watashi no kao o mire-ba nai-ta.

"The neighbor's kid cried whenever she saw my face."

Haru ni nar-eba sakura ga saku.

"Whenever spring comes, cherry flowers bloom."

Hito o koros-eba keimysho iki-da.

"It is natural for a murderer to go to jail."

To spans the domains of A, B, and C.

A. Past events expressing a "when" meaning.

Taroo wa heya ni hairu-to suguni konpyuutaa geemu o hajime-ta.

"Taroo started a computer game as soon as he entered the room."

Hitori de kooen o aruite-iru-to, futo atarashii kangae ga ukan-da.

"When I was walking in a park alone, I came up with a new idea."

B. Past or nonpast events expressing an "if" meaning.

Kore o nom-anai-to yoku nar-anai.

"If you do not take this, you will not get better."

Ashita hareru-to iinaa.

"I hope it will be sunny tomorrow."

Kinoo hareru-to yokat-ta noni.

"I wish it was sunny yesterday."

C. Past or nonpast habitual events expressing a "when/whenever/ every time" meaning. A "when" interpretation here is a general event interpretation indicating a natural course of events.

Haru ni naru-to tsubame ga kite, su o tsukuru.

"When spring comes, swallows come and nest."

Anoko wa watashi no kao miru-to kimatte naku.

"That child cries whenever s/he sees my face."

Based on the illustration, modality subcategorized in this book is best seen in the domain marked as "B" in figure (91), where all three conditionals express hypothetical notions, where hypotheticality and epistemic possibility merge. Thus modality communicates the notions of semelfactivity and indeterminability that enable conditionals to express hypotheticality.

CHAPTER 6

Conclusions

In mathematics, a positive multiplied by a positive begets a positive, as does a negative multiplied by a negative. Correspondingly, in simple propositional logic, the operation of negation is to reverse the truth value of a proposition, as indicated in the basic logical axiom: $\sim \sim p = p$ (where $\sim$ is the symbol for negation and p, proposition). Nevertheless, these theoretical symbols are not consistently valid when "real life" is taken into account. As Givon (1975) points out, affirmatives and their corresponding negatives in natural language differ not only in their truth value, but also in their discourse presuppositions. The discourse context—or pragmatic presupposition—in which negative propositions are used in natural language is richer than that of their corresponding affirmative propositions. Negation has a pragmatic component that cannot be predicted from logic. A similar blind spot in the relationship between logic and natural language can be found in conditional constructions in which the logical relationship between the antecedent and the consequent is the same as that given for material implication (if p, then q). For example, the sentence "If the Nile runs through Egypt, then the earth is round" is logically true since the antecedent and the consequent are both true. In material implication, the conditional is false only when the antecedent is true, and the consequent is false regardless of the causal relation. In natural language, however, sentences like this are not necessarily valid.

In the study of modality in western languages, notions of necessity and possibility came to be treated as objects of epistemic modes (human knowledge and belief) rather than alethic modes (logical truth of the proposition). Nevertheless, notions of logical necessity and possibility, like logical negation, sometimes do not express meanings which reflect "real life" situations. For example, "It is not possible that it is not raining in Tokyo" can be logically assumed to express the same meaning as "It is necessary that it is raining in Tokyo." Logical necessity here, however, does not fully reflect the conjectural judgment behind the expression. Sweetser (1990, p. 113) also states that natural-language speakers require a connection between the clauses, and conditionals are used to refer to "related" things. It was the recognition of these logical shortcomings that led language researchers like Lyons, Palmer, and Coates to concentrate on semantic considerations of language, recognizing that language reflects human knowledge and belief, neither of which can be precisely presented mathematically. Moreover, it is clearly the case that modality is concerned with human knowledge and belief; however, the study of modality tends to be chaotic, as seen in Palmer's work and in recent Japanese studies, since knowledge and belief at root communicate a speaker's psychological attitude, which itself is a more problematic concept still. Hence, the study of modality can be approached from any number of viewpoints, ranging from the study of discrete lexical items to the study of prosody.

In Japanese, the study of modality began when the grammatical categories of propositional content and modal content were established. Propositional content encompasses that which expresses objectified notions, and modal content encompasses that which expresses a speaker's psychological attitude. The line between these is drawn by a tense marker. Although this division cannot rigidly be determined as a syntactic phenomenon, the semantic function of the constituents in modal content serves to clarify the distinction, which is one of the unique characteristics of the Japanese language and plays a significant role in deciding the central notion of modality.

By taking advantage of previous research, the investigation herein refocused modality in order to establish a more effective approach to its study. This was done first by reexamining and applying the notions of epistemic and deontic necessity and possibility, which are the central notions of modality in western language, to modal content. These notions are expressed in modal content by suppositionals and evidentials that are used to represent the degree of speaker confidence in the truth or realization of a proposition.

The notions of necessity and possibility, however, do not exist in two completely different worlds; rather, they point to polar opposites, one marking the highest degree of necessity (the highest degree of actuality of a proposition) and the other the lowest degree of possibility (the lowest degree of actuality of a proposition). This was proven by examining both the semantic and syntactic behavior of suppositionals and evidentials. In these investigations it was demonstrated that the higher the degree to which modals express the notion of necessity, the higher the degree of potentiality that the modals behave as a constituent of the proposition. This observation was supported most clearly by the behavior of the suppositional *hazu*, which can partake in the past tense or negative form, for example, *hazu-dat-ta* and *hazu-ja-nai/hazu wa nai,* respectively. These forms of *hazu* no longer represent speaker supposition, but statements. Even when *hazu* is used in the affirmative, nonpast tense form, it can be perceived as expressing a pre-planned notion. This is discerned in the survey response, for example, to sentences (64)a, *Kuruma o kau hazu-da ga kaw-anai,* "I'm sure that I will buy a car, but I will not (buy a car)" and (66)a, *Taroo wa hanako ni au hazu daroo,* "It is probably the case that Taro must meet Hanako." Some respondents perceived this *hazu* as *kotoni-natte-iru* "it has been decided," and such interpretation perhaps caused them to allow the co-occurrence of a counter situation and another modal, respectively. Thus *hazu,* which expresses a high degree of speaker conviction and therefore a high degree of necessity, can vacillate between the domain of propositional content and modal content.

A similar observation was made regarding the behavior of one of the evidentials—visual *soo.* In its past tense and negative forms, *soo* does not function as a modal but constitutes part of a proposition. The difference between *hazu* and *soo* stems from the notions of supposition and evidence, where the former does not necessarily involve visual or sensory testimony, and the latter largely relies on such sources. Witnessing the actual occurrence of an event does not allow the speaker to evaluate or reflect upon the entire event externally and objectively, therefore prohibiting the speaker from speculating on the realization of an event. Hence, the proposition regarding the occurrence of the event cannot accompany another modal. This was the subtle difference detected in the appropriateness acknowledged in sentences (66)a, *Taroo wa Hanako ni au hazu daroo* "It is probably the case that Taro must meet Hanako," and (66)e, *Ame ga furi-soo kamoshire-nai*, "It may be the case that it looks like it is about to rain" in chapter 3, and one of the keys in deciding that *soo* expresses the higher degree of necessity as viewed by the speaker.

The highest degree of modality (i.e., the lowest degree of possibility) is found in the use of the suppositional *kamoshire-nai*, which functions as a modal even in the past tense form (*kamoshire-nakat-ta*). In this case, a speaker simply supposes the truth of the realization of an event described in the proposition from a remote situation. There is no guarantee or presentable evidence to support his/her statement, and the expressed situation is merely conjectural, where the truth of the proposition is indeterminable. The center feature of modality, therefore, is its relation to possible/nonactual world situations. These situations are found at the lowest degree of epistemic possibility, and the notions of necessity and possibility are expressed by both the semantic interpretation of modals and their syntactic behavior. When a modal expresses a high degree of necessity, its past tense and negative forms certainly become constituents of the proposition and have the ability to invite another modal. On the other hand, a modal expressing a low degree of possibility, even in the past tense form, can express uncertainty of the situation as viewed by the speaker.

It was also found that among the modals, *daroo* singularly functions as a pure modal, expressing a speaker's judgment at the time of speech. This observation is valid since *daroo* does not occur in the negative, past tense form, and it does not allow expression of another individual's supposition. This was exemplified by the inappropriateness of the co-occurrence of hearsay *soo*, as in *Saifu wa hikidashi ni aru daroo soo-da*, "I hear that I think it is probably the case that the wallet is in the drawer." It is interesting to note that this pure modal *daroo* covers the widest range of modality.

Through careful scrutiny of the modals, it was shown that epistemic possibility is the central notion of modality. Deontic modality is, by contrast, mainly expressed by the notion of necessity, which itself is syntactically in the domain of propositional content. Although the permissive expression *te-mo ii*, "it is okay if you do ~," conveys the notion of possibility, it also lies in the domain of propositional content. Deontic modality is expressed by a speaker's attitude toward an individual, while epistemic modality is expressed by a speaker's judgment toward a proposition. The common point is that both modalities are based on a speaker's knowledge and belief, although to varying degrees when the notions of possibility and necessity are considered. The lower the degree of possibility that a proposition is true or realized, the greater the degree of modality. Modality then can also be approached as present when the truth and/or the realization of a proposition is least possible. This places modality in the domain of possible/nonactual worlds, and suggests a new subcategory of modality within the larger definition of a speaker's psychological attitude. Therefore modality herein particularly refers to possible/nonactual worlds.

The investigation of modals also revealed that possible/nonactual world situations are deeply entwined with the notion of indeterminability in that the truth or realization of a proposition is related to the speaker's not possessing firm evidence about it. While propositions with no practical connection are acceptable in theoretical logic studies, as in examples involving material implication such as "if the Nile runs through Egypt, then the earth is round," the use of practical language is based on human knowledge and belief. However, as expressed in natural language, human knowledge and belief do not necessarily have universal validity. Modality can therefore be said to be a completely speaker-oriented concept. Each individual's utterances are based on his/her own knowledge and beliefs, and these may appear to be subjective to other individuals. Even so, if one has what one perceives to be a solid basis for a judgment, then the judgment is objective to the speaker him/herself.

The notion of indeterminability is often triggered by ungeneralized events, future events not under the speaker's control, and natural phenomena. Determinability, on the other hand, can be triggered by a natural course of events, general events, a speaker's self-knowledge (volitionally controllable events), and past events. The notions of indeterminability and determinability of a proposition are respectively associated with hypotheticality and nonhypotheticality. Indeterminable possible/nonactual world situations are associated with hypothetical worlds in that the proposition does not possess truth value, hence, possible/nonactual world situations relate to hypothetical situations.

Conditional sentences represent hypothetical situations. As evidenced by the fact that they take a modal, conditional sentences fall under propositional content, as in *Okane ga are-ba kuruma ga ka-eru kamoshire-nai* "It may be that if I have money I can buy a car." By examining the conditionals *to, ba,* and *tara,* it was observed that the greater the degree to which a situation is judged to be indeterminable from the speaker's viewpoint, the greater the degree to which the situation becomes hypothetical, that is, the truth value of the situation becomes less distinct. This means that the interpretation of the notions of "when" or "if" is the same as that of "necessity" and "possibility," respectively, and that the notions of "when" and "if" are located at two opposite ends of a bipolar continuum. Japanese happens to express these notions by various conditional forms.

Of all the conditionals, it was found that *ba* sentences express the highest degree of hypotheticality. This is due to the fact that even though *ba* sentences are used in past contexts, they do not express particular past events, but only past habitual/repeated events or counterfactual events (indeterminable situations). A determinable, controllable event in the antecedent excludes hypotheticality, and therefore this type of event in a *ba* construction is irreconcilable with a nonactual situation in the consequent. Volitional expressions are not-yet-realized situations, and therefore are in the domain of nonactual worlds. These expressions are inappropriate when a predicate containing a *ba* construction is determinable, as was seen, for example, in the inappropriateness of a sentence such as *Kono shigoto o oere-ba sono shigoto o suru,* "If I finish this job, I will do that job." In this sentence, the verb *oeru* "to finish" is used to describe a determinable situation due to the speaker's knowing his/her own commitment to the event. When a *ba* construction involves a nonvolitional verb, on the other hand, the sentence is appropriate, as, for example, in *Kono shigoto ga owar-eba sono shigoto o suru,* "If this job is finished I will do that job." Thus determinability and controllability of a predicate in a *ba* construction are decisive in what type of event can be predicated in the consequent. A close examination of conditional sentences also indicated that the notions of indeterminability and hypotheticality are deeply related to the speaker's volitional control of a situation. When a situation is indeterminable, it is not only hypothetical, but also beyond a speaker's control, as with natural phenomena, unaccusative events, and another individuals' future activities—all of which invite hypotheticality and indeterminability. Therefore, possible/nonactual worlds are concerned with situations where a speaker cannot exert volitional control over the realization of an event.

When modality is taken into account in possible/nonactual world situations, a lack of speaker control creates a spatiotemporal and psychological distance between the situation and the speaker. In this sense, situations that are spatiotemporally and psychologically close to the speaker are outside the domain of modality. This was illustrated in various ways. For example, the hypothetical *ba* construction cannot be used to describe a situation occurring or about to occur in the immediate presence of the speaker. This was specifically exemplified by the inappropriateness of the sentence, *Abunai! otos-eba wareru* "Watch out! If you drop it, it will break." The immediacy of this type of event contradicts the indeterminable, highly hypothetical situation expressed by *ba* sentences. In order for a *ba* sentence to be appropriate, the situation has to be one in which there is sufficient spatiotemporal distance for the speaker to process the causal relation. One such example was seen in the appropriateness of the sentence, *Kono kusuri o noma-nake-reba yoku nar-anai,* "If you do not take this medicine, you will not get better." Such a sentence describes not an immediate event, but one for which the speaker is free to process pertinent considerations both mentally and physically. The negative suffix *nai, in nom-anake-reba,* "if you don't take," which expresses a static state of a nonspecific duration, supports the validity of such an observation.

Another example of the necessary existence of a spatiotemporal distance between a speaker and a situation was seen in the behavior of modals. A slight difference was detected between the various uses of visual *soo* "it appears to be," depending on the type of accompanying predicate. A nonhomogeneous predicate with *soo,* such as *ochi-soo,* "it looks like it is about to fall," expresses a higher degree of actuality (necessity) than a homogeneous stative predicate, such as *taka-soo* "it looks expensive." In a situation as immediate as that in *ochi-soo,* there is no need for one to observe and process the possibility of an event's occurrence as it is so close to entering the domain of an actual event—"to fall." A stative predicate with *soo,* such as *taka-soo,* however, expresses a situation where the spatiotemporal distance allows the speaker to process his/her own judgment—that is, the speaker speculates on the price of the item s/he is looking at and comes to the conclusion that it is expensive. Based on these observations, it is concluded that volitionally uncontrollable, indeterminable situations are more consistent with stative, static situations than they are with dynamic situations. The relationship between uncontrollable, indeterminable and stative, static situations is reflected by the role stative predicates play in conditional sentences.

Stativity is expressed by the aspectual affix *te-iru* (resultative state) and by adjectives. The negative suffix *nai* is morphologically an adjective. Therefore, *te-iru* and the negative suffix *nai* share a common feature in that they are both homogeneous predicates, and are thus in the same aspectual category. Close scrutiny of this homogeneous category, however, revealed that it included two types of situations: purely static situations that do not require energy input; and progressive/iterative situations that require constant energy input to maintain them. *Te-iru* can express either of these as well as temporally cyclical situations that are variously transformed from dynamic situations into static situations or vice versa. This is seen, for example, in *Asagohan wa moo tabete-iru* "I have eaten breakfast" and *Ima bangohan o tabete-iru* "I'm eating dinner now." *Te-iru* vacillates between the static and dynamic.

Although *te-iru* is used to express actual world situations in simple sentences, and is consequently outside of the domain of modality, the static form is found to play a meaningful role in increasing the degree of hypotheticality in conditional *ba* sentences. This is particularly exemplified in past contexts in which *te-iru* is used to distinguish a counterfactual event (hypothetical event) from a past habitual event. Representative examples are *Ame no hi ni ik-eba mise wa suite-i-ta,* "Every time we went to the store on a rainy day, it was not crowded/If I went to the store on a rainy day, the store would not have been crowded," which can be interpreted either as a past habitual event or a counterfactual event, and *Ame no hi ni itte-ire-ba mise wa suite-i-ta,* "If I went to the store on a rainy day, the store would not have been crowded," which includes *te-iru* and can only receive a counterfactual interpretation. Reflecting a situation after the introduction of energy, this *te-iru,* used in *ba* sentences, expresses only a resultative state, and the counterfactual event is in the domain of a possible/nonactual world that the speaker's volitional control cannot manipulate. Contrarily, *te-iru* used in a dynamic situation, which is under a speaker's volitional control, does not garner a hypothetical interpretation, but instead receives a progressive/repetitive interpretation that is in the domain of the actual world.

Dynamic worlds described by *te-iru* can be interpreted as habitual. For example, *ima tabete-iru*, "I am eating now," can be a one-time progressive event, while *mainichi tabete-iru* "I am eating everyday" can be interpreted as expressing habitual activity. Thus, progressivity and habituality are not two different worlds, but refer to the same world viewed as a continuum—one extreme corresponding to a single progressive activity, and the other to habitual activity. The transition from a single progressive activity to a habitual activity can be envisioned in analogy to the start-up of an electric fan. As the fan begins to gain speed, the outline of the individual fins disappears and is replaced by a seemingly solid circle. So too when a single progressive activity is repeated: it becomes habitual. Each single activity is less distinct than the habituality itself, and the notion of habituality approaches that of stativity. In this sense, habituality and stativity share similar characteristics, with one significant difference in that habituality requires energy input whereas stativity does not. This is the key to distinguishing progressivity/iterativity, which is concerned with actual world situations, from stativity. Takubo (1993) pointed out that Jacobsen's statement (1990a) regarding the cognitive relationship between stativity and nonactual world situations is seen only in conditional sentences (subordinate clauses, not main clauses). It should be noted, however, that the role of stativity relating to nonactual world situations is not merely seen in conditional sentences, but also in various constructions supporting Jacobsen's premise. It is through this variety of subordinate clauses that stativity can be seen to increase the degree of hypotheticality in sentences that express modality.

The concepts of "dynamic" and "stative" can be analogically applied to the notions of "necessity" and "possibility," which in turn communicate the notions of "whenever/when" and "if," respectively. These notions are not in two unrelated worlds; rather, they are situated on a continuum relating one to the other. As the degree of "necessity" and "possibility" is examined and defined by the semantic investigation of modals, the degree is also identified by the use of conditionals forms—one end expressing a "whenever/when" meaning and the other end "if." The reason *ba,* for example, can express "whenever" and "if" is that these notions are captured on the same continuum, varying the degree of the possibility of the truth of a proposition.

The negative suffix *nai* is morphologically identical to an adjective, and it possesses characteristics similar to *te-iru*—for example, in terms of aspect and the vacillation between dynamic and static worlds; hence, it can express stativity, and it might be expected that *nai* would be related to possible/nonactual world situations and play a meaningful role in increasing the degree of hypotheticality. Investigation of *nai,* however, unequivocally revealed that it is substantially affected by its corresponding affirmative form, as exemplified in the fact that the negative form of nonstative volitional predicates can co-occur with *te-iru,* whereas stative predicates cannot. Such indicates that volitional (inhibitional) energy can be introduced to a negative situation, as seen in the example sentence *Kodomo ga nak-anaide-iru* "The child is trying not to cry." Therefore, *nai,* is largely ineffective in converting a nonstative predicate to a stative predicate when the corresponding affirmative form is a nonstative predicate.

This finding was further confirmed through examination of the behavior of *nai* in conditional *ba* sentences, where the inappropriateness of determinable, volitionally controllable predicates in an antecedent that is connected to a consequent expressing a nonactual situation cannot be rectified merely by converting the predicate in the antecedent into a negative form of itself. This was evident in the inappropriateness of the example sentences *Kono hon o yom-eba sono hon o yomu,* "If I read this book, I will read that book" and *Kono hon o yomanake-reba sono hon o yomu,* "If I don't read this book, I will read that book." The features of the original affirmative form dominate the behavior of the negative suffix *nai* in conditional *ba* sentences as well. Consequently, when the corresponding affirmative is a stative predicate, such as an adjective, both the affirmative and negative forms of the predicate become appropriate for use with *ba* as a way of expressing a high degree of hypotheticality.

In both internal and external negation structures of simple sentences, it was found that *nai* can be used to express nonactual world situations when the focus of negation is not on the element to which *nai* is attached. For example, the actual event in the sentence, *Sonoko wa yooka ni umare-nakat-ta,* "The child was not born on the eighth of the month" is the birth of the child. Here *umare-nakat-ta,* "was not born," refers to a nonactual world situation; nevertheless, such situations expressed by negation cannot be viewed in the same manner as the nonactual world situations expressed by modals or conditional sentences. That is, nonactual world situations expressed by negative sentences are not indeterminable hypothetical situations as viewed by the speaker, but are based on actual world situations known to the speaker or that can be figured out by the speaker. This is due to the pragmatic aspects of negative sentences, namely that "a negative speech act is shown to be presuppositionally more marked than that of affirmatives" (Givon 1975). The examination of the focus of negation using external negation structures also revealed that suppositionals, evidentials, and explanatory modals used in the external negation structure *-no-da* and *wake-da* have a very different function in that suppositionals and evidentials are in nonactual world domains while explanatory modals are in the actual world domain, even though they are all in the domain of modal content. Thus, the aspectual forms *te-iru* and *nai* are both concerned with actual world situations and are outside the domain of modality as described herein.

What we have seen, however, is that pure stative predicates expressed by the negative suffix *nai* have the same function as the resultative *te-iru.* Both are used in situations where a spatiotemporal distance is created between a speaker and a situation over which the speaker has no volitional control. Although the situations expressed by the predicate can be recognized as actual world situations, when stative predicates play a role in possible/nonactual world situations, they are responsible for expressing the highest degree of epistemic modality. The highest degree of modality is thus expressed in possible/nonactual, indeterminable, hypothetical worlds where human volitional control cannot operate. These notions are all located toward one end of a continuum, with the opposite notions on the opposite end. Modals and conditional *ba* sentences thus exemplify this concept in that they express possible/nonactual, indeterminable, hypothetical notions as viewed by the speaker.

The investigations presented herein also pointed out that modals in modal content have different functions, particularly regarding the distinction between suppositionals and evidentials, and explanatory modals. The latter are concerned with a speaker's pregathered knowledge of the world—that is, they reflect actual world situations, just like propositions described by aspectual forms. This suggests that yet another entrance to the study of modality is plausible, approached from the viewpoint of actual world situations.

Since the general definition of modality as "a speaker's psychological attitude" is such a broad notion, it is unavoidable that numerous approaches to the study of modality have developed. Under such circumstances, it is sensible to try to identify a thread common to each approach. By tracing the origins of the study of modality, this book revealed such a thread. Modality was investigated from the viewpoints of the notions of necessity and possibility, which themselves originated in studies of modality in western languages. Hence, modality can be viewed as the degree of the possibility of a proposition's truth or realization as viewed by the speaker. This definition adds practical and applicable perspectives to the study of Japanese modality in that propositions also become objects of that study, namely, conditional sentences and aspectual categories. Applying these notions to the Japanese language demonstrates a novel entrance to the study of modality through investigation of possible/nonactual worlds, revealing a common thread for others to follow.

Bibliography

Akatsuka, Noriko. 1983. Conditionals. *Papers in Japanese Linguistics* 9: 1–34.

———. 1986. Conditionals are Discourse-Bound. In *On Conditionals,* edited by E. Traugott et al. Cambridge: Cambridge University Press.

Alfonso, Anthony. 1966. *Japanese Language Patterns*. Tokyo: Sophia University Press.

Austin, John. 1962. *How to Do Things with Words*. Cambridge, Mass.: Harvard University Press.

Barnes, Janet. 1984. Evidentials in the Tuyuca Verb. *International Journal of American Linguistics* 50: 255–71.

Bybee, Joan, and S. Fleishman, eds. 1995. *Modality in Grammar and Discourse*. Philadelphia: John Benjamins.

Bybee, Joan, and W. Pagliuca. 1994. *The Evolution of Grammar*. Chicago: University of Chicago Press.

Carlson, Lauri. 1981. Aspect and Quantification. In *Syntax and Semantics 14,* edited by P. Tedeschi and A. Zaenen. New York: Academic.

Chierchia, Gennaro, and S. McConnell-Ginet. 1992. *Meaning and Grammar*. Cambridge, Mass.: MIT Press.

Chomsky, Noam. 1965. *Aspects of the Theory of Syntax*. Cambridge, Mass.: MIT Press.

Coates, Jennifer. 1983. *The Semantics of the Modal Auxiliaries*. London: Croom Helm.

Comrie, Bernard. 1976. *Aspect*. Cambridge: Cambridge University Press.

Dahl, Osten. 1981. On the Definition of the Telic-Atelic (Bounded-Non-bounded) Distinction. In *Syntax and Semantics 14*, edited by P. Tedeschi and A. Zaenen. New York: Academic.

Dancygier, Barbara. 1998. *Conditionals and Prediction*. New York: Cambridge University Press.

Dowty, David. 1975. The Stative in the Progressive and Other Essence/ Accident Contrasts. *Linguistic Inquiry* 6: 465–74.

———. 1979. *Word Meaning and Montague Grammar*. Dordrecht: D. Reidel.

Eda Sumire. 1991. Fukugōji ni yoru Jōken Hyōgen I: To naruto no Imi to Kinō. *Nihongo Kyōiku* 75: 153–63.

Fauconnier, Gilles. 1994. *Mental Spaces: Aspects of Meaning Construction in Natural Language*. New York: Cambridge University Press.

Fillmore, Charles. 1968. The Case for Case. In *Universals in Linguistic Theory*, edited by E. Back and R. Harms. New York: Holt, Rinehart and Winston.

Galton, Antony. 1984. *The Logic of Aspect*. Oxford: Clarendon.

Geis, Michael, and A. Zwicky. 1971. On Invited Inference. *Linguistic Inquiry* 2: 561–66.

Givon, Talmy. 1975. Negation in Language: Pragmatics, Function, Ontology. *Working Papers on Language Universals [WPLU]* 18: 59–115.

Gruber, Jeffrey. 1967. Look and See. *Language* 43: 937–47.

Hashimoto Shinkichi. 1948. *Kokugo-hō Kenkyū*. Tokyo: Iwanami Shoten.

———. 1969. *Joshi, Jo-Dōshi no Kenkyū*. Tokyo: Iwanami Shoten.

Hasunuma Akiko. 1993. Tara to To no Jijitsuteki Yōhō o Megutte. In *Nihongo no Jōken Hyōgen*. Tokyo: Kurosio.

Hayatsu Emiko. 1998. Rashii and Yōda. *Nihongo-gaku* 7: 46–61.

Hinds, John, and Wako Tawa. 1975. Conditions on Conditionals in Japanese. *Papers in Japanese Linguistics* 4: 3–12.

Hintikka, Jaakko, ed. 1969. *Models for Modalities*. Dordrecht: D. Reidel.

Hoeksema, Jack. 1985. *Categorial Morphology*. New York: Garland.

Inaba Midori. 1991. Nihongo Jōken-bun no Imi Ryōiki to Chūkan Gengo Kōzō. *Nihongo Kyōiku* 75: 153–63.

Ishigami Teruo. 1990. Hitē to Kōbun. *Nihongo-gaku* 9: 57–67.

Jackendoff, Ray. 1990. *Semantic Structures*. Cambridge, Mass.: MIT Press.

Jacobsen, Wesley. 1982. Vendler's Verb Classes and the Aspectual Character of Japanese Te-iru. *Berkeley Linguistics Society* 8: 373–83.

———. 1984. Lexical Aspect in Japanese. *Papers from the Parasession on Lexical Semantics*, 150–61. Chicago: Chicago Linguistics Society.

———. 1990a. Jōken-bun ni Okeru Kanrensē ni Tsuite. *Nihongo-gaku* 9: 93–108.

———. 1990b. The Mutifaceted Character of Stativity in Japanese. In *On Japanese and How to Teach It*, edited by O. Kamada and W. Jacobsen. Tokyo: The Japan Times.

———. 1992a. Are Conditionals Topics? The Japanese Case. In *The Joy of Grammar*, edited by D. Brentari et al. New York: John Benjamins.

———. 1992b. *The Transitive Structure of Events in Japanese*. Tokyo: Kurosio.

———. 1997. Agentivity and Aspect in Japanese: A Functional Perspective. In *Directions in Functional Linguistics*, edited by A. Kamio. Amsterdam and Philadelphia: John Benjamins.

Jespersen, Otto. 1924. *The Philosophy of Grammar*. London: Allen and Unwin.

Johnson, Yuki. 1990. One Aspect of Japanese Modality: The Use of Sō, Yō (Mitai), and Rashii. *International Christian University Language Research Bulletin* 5.1: 1–20.

Jorden, Eleanor, and Mari Noda. 1990. *Japanese: The Spoken Language, Part 3*. New Haven and London: Yale University Press.

Kageyama Taro. 1996. *Dōshi Imi-ron*. Tokyo: Kurosio.

Kamio, Akio. 1990. *Jōhō no Nawabari Riron*. Tokyo: Taishukan Shoten.

———. 1994. The Theory of Territory of Information: The Case of Japanese. *Journal of Pragmatics* 21: 67–100.

———. 1997a. Evidentiality and Some Discourse Characteristics in Japanese. In *Directions in Functional Linguistics*, edited by A. Kamio. Amsterdam and Philadelphia: John Benjamins.

———. 1997b. *Theory of Territory of Information*. Amsterdam and Philadelphia: John Benjamins.

Karttunen, Lauri. 1972. Possible and Must. In *Syntax and Semantics 1*, edited by J. Kimball. New York: Academic.

Katoo, Yasuhiko. 1985. Negative Sentences in Japanese. In *Sophia Linguistica 19*. Tokyo: Sophia University Press.

Kawabata, Yasunari. 1981. *Snow Country*. Translated by Edward Seidensticker. 1957. New York: Alfred A. Knopf, Inc.

Kawaguchi, Sachiko. 1984. To, Ba, Tara—Danwa ni okeru Sentaku Yōin o Motomete. *Waseda University Research Bulletin* 28: 1–18.

Kenny, Patrick. 1963. *Action, Emotion and Will*. London: Routledge and Kegan Paul.

Kindaichi Haruhiko. 1976a. Kokugo Dōshi no Ichibunrui. In *Nihongo Dōshi no Asupekuto*, edited by Kindaichi Haruhiko. Tokyo: Mugi Shobō.

———. 1976b. Nihongo Dōshi no Tensu to Asupekuto. In *Nihongo Dōshi no Asupekuto*, edited by Kindaichi Haruhiko. Tokyo: Mugi Shobō.

Kinoshita, Rika. 1997. Hazu-da no Imi Bunseki. *Nihongo Kyōiku* 92: 165–76.

Kitagawa, Chisato, and A. Iguchi. 1988. *Japanese for Foreigners: Jodōshi*. Tokyo: Aratake Shuppan.

Koide, Keeichi, Noriko Komatsu, and Izumi Saita. 1981. To, Ba, Tara. *Stanford Inter University Center for Japanese Language Studies Bulletin* 4: 30–66.

Kripke, Saul. 1959. A Completeness Theory in Modal Logic. *Journal of Symbolic Logic* 24: 1–14.

Kuno Susumu. 1973. *The Structure of the Japanese Language*. Cambridge, Mass.: MIT Press.

———. 1983. *Shin Nihon Bunpō Kenkyū*. Tokyo: Taishūkan Shoten.

Lascarides, A. 1988. A Formal Semantics of the Progressive. Ph.D. dissertation. University of Edinburgh.

Levin, Beth, and Malka Rappaport Havov. 1995. *Unaccusativity*. Cambridge, Mass.: MIT Press.

Lewis, David. 1973. *Counterfactuals*. Oxford: Blackwell.

———. 1991. Counterfactual Dependence and Time Arrow. In *Conditionals*, edited by F. Jackson. London: Oxford University Press.

Lyons, John. 1977. *Semantics*. Cambridge: Cambridge University Press.

Machida Ken. 1989. *Nihongo no Jisei to Asupekuto*. Tokyo: ALC Press.

Makino, Seiichi, and M. Tsutsui. 1989. *A Dictionary of Japanese Grammar*. Tokyo: The Japan Times.

Masuoka Takashi. 1989a. *Mēdai no Bunpō*. Tokyo: Kurosio.

———. 1989b. Modality. In *Nihongo to Nihongo Kyōiku 12*, edited by Kondō T. Tokyo: Meiji Shoin.

———. 1995. *Nihongo no Jōken Hyōgen ni Tsuite*, edited by T. Masuoka. In *Nihongo no Jōken Hyōgen.* Tokyo: Kurosio.

Matsushita Daizaburō. 1901. *Nihon Zokugo Bunten*. Tokyo: Seibundō Shoten.

———. 1930. *Hyōjun Nihon Bunpō*. Tokyo: Kigensha.

Maynard, Senko. 1993. *Discourse Modality: Subjectivity, Emotion, and Voice in Japanese*. New York: John Benjamins.

McGloin, Naomi Hanaoka. 1976a. Negation. In *Syntax and Semantics 5*, edited by M. Shibatani. New York: Academic.

———. 1976b. The Speaker's Attitude and the Conditionals To, Ba, and Tara. *Papers in Japanese Linguistics* 5: 181–92.

———. 1990. Hitē Hyōgen no Nichi-Ee Taishō Hyōgen. *Nihongo-gaku* 9: 68–75.

Mikami Akira. 1963. *Nihongo no Kōbun*. Tokyo: Kurosio.

Miyagawa, Shigeru. 1989. *Structure and Case Marking in Japanese*. New York: Academic.

Miyata Kōichi. 1948. *Nihongo Bunpō no Rinkaku*. Tokyo: Sansei-dō.

Moens, M., and M. Steedman. 1987. Temporal Ontology in Natural Language. *Proceedings of the 25th Annual Meeting of the Association for Compositional Linguistics*: 1–7.

Morita Yoshiyuki. *Kiso Nihongo*. Tokyo: Kadokawa Shoten.

Moriyama Takurō. 1988. *Nihongo Dōshi Jutsugo-bun no Kenkyū*. Tokyo: Meiji Shoin.

———. 1989. Ninshiki no Mūdo to Sono Shūhen. In *Nihongo no Modality*, edited by Nitta Yoshio and Masuoka Takashi. Tokyo: Kurosio.

Mourelatos, D. 1978. Events, Process, and States. *Linguistics and Philosophy* 2: 415–34.

Nakau, Minoru. 1979. Modality to Mēdai. In *Eego to Nihongo to*. Tokyo: Kurosio.

Nitta Yoshio. 1986. Genshō-byōsha-bun o Megutte. *Nihongo-gaku* 5: 56–69.

———. 1989. Gendai Nihongo-bun no Modality no Taikē to Kōzō. In *Nihongo no Modality*, edited by Nitta Yoshio and Masuoka Takashi. Tokyo: Kurosio.

Ōkubo Tadatoshi. 1968. *Nihon Bunpō Chinjutsu Ron*. Tokyo: Meiji Shoin.

Okuda Yasuo. 1984. *Kotoba no Kenkyū Josetsu*. Tokyo: Mugi Shobō.

Ōta Akira. 1980. *Hitē no Imi: Imi-ron Josetsu*. Tokyo: Taishūkan Shoten.

Ōtsuki Fumihiko. 1904. Gohō-Shinan. In *Genkai*, edited by Ōtsuki Fumihiko. Tokyo: Rokugō Kan.

Palmer, Frank. 1977. Modals and Actuality. *Journal of Linguistics* 13: 1–23.

———. 1979. Why Auxiliaries are Not Main Verbs. *Lingua* 47: 1–25.

———. 1986. *Mood and Modality*. Cambridge: Cambridge University Press.

———. 1990. *Modality and the English Modals*. New York: Longman.

Perlmutter, David. 1978. Inversonal Passives and the Unaccusative Hypothesis. *Berkeley Linguistics Society* 4: 157–89.

Poutsma, Hendrik. 1926. *A Grammar of Late Modern English*. 2nd ed. Gloningen: P. Noordhoff.

Radford, Andrew. 1988. *Transformational Grammar*. Cambridge: Cambridge University Press.

Recher, Nicholas. 1968. *Topics in Philosophical Logic*. Amsterdam: D. Reidel.

Reilly, Judy. 1986. The Acquisition of Temporals and Conditionals. In *On Conditionals*, edited by E. Traugott et al. Cambridge: Cambridge University Press.

Reinhart, Tanya. 1983. *Anaphora and Semantic Interpretation*. London: Croom Helm.

Ross, John. 1976. Auxiliaries as Main Verbs. In *Studies in Philosophical Linguistics I*, edited by W. Todd. Evanston: Great Expectations.

Sagawa Makoto. 1976. Nihongo no Hitē no Han'i. *Gengo Kenkyū* 70: 1–22.

Sakuma Hajime. 1936. *Gendai Nihongo no Hyogen to Gohō*. Tokyo: Kōseikaku.

Schachter, Jacquelyn. 1971. "Presupposition and Counterfactual Conditional Sentence." Ph.D. dissertation. University of California, Los Angeles.

Soga, Matsuo. 1983. *Tense and Aspect in Modern Colloquial Japanese*. Vancouver: University of British Columbia Press.

Stalnaker, Roberts. 1984. *Inquiry*. Cambridge, Mass.: MIT Press.

Suga Kazuyoshi and Emiko Haytsu. 1995. *Dōshi no Jita*. Tokyo: Hitsuji Shobō.

Sugifuji Miyoko. 1992. Prosody to wa Nani ka. *Gengo* 21: 16–21.

Suzuki, Hideo. 1990. Setsuji, Settogo, Setsubigo. *Nihongo-gaku* 9: 52–56.

Suzuki, Shigeyuki. 1972. *Nihongo Bunpō Kētai Ron*. Tokyo: Mugi Shobō.

Sweetser, Eve. 1990. *From Ethmology to Pragmatics*. Cambridge: Cambridge University Press.

Takahashi, Tarō. 1969. Sugata to Mokuromi. In *Nihongo Dōshi no Asupekuto*, edited by H. Kindaichi. Tokyo: Mugi Shobō.

Takubo, Yuki nori. 1993. Danwa Kanri kara Mita Nihongo no Hanjijitsu Jōkenbun. In *Nihongo no Jōken Hyōgen*. Tokyo: Kurosio.

Taylor, Barry. 1977. Tense and Continuity. *Linguistics and Philosophy* 1: 199–220.

Tenny, Carol. 1987. "Grammaticalizing Aspect and Affectedness." Ph.D. dissertation. MIT.

———. 1994. *Aspectual Roles and the Syntax-Semantics Interface*. Boston: Kluwer Academic.

Teramura Hideo. 1971. Ta no Imi to Kinō. In *Gengogaku to Nihongo Mondai*. Tokyo: Kurosio.

———. 1979. Muudo no Keeshiki to Hitee. In *Eego to Nihongo to*. Tokyo: Kurosio.

———. 1982. Tensu, Asupekuto no Koto teki Sokumen to Mūdo teki Sokumen. *Nihongo Kyōiku* 57: 4–16.

———. 1984. *Nihongo no Shintakusu to Imi II*. Tokyo: Kurosio.

Tokieda Motoki. 1950. *Nihon Bunpō Kōgo-hen*. Tokyo: Iwanami Shoten.

———. 1955. *Kokugo-gaku Genron*. Tokyo: Iwanami Shoten.

Toyota Toyoko. 1979. Hakken no To. *Nihongo Kyōiku* 36: 91–105.

Tsujimura, Natsuko. 1991. On the Semantic Properties of Unaccusativity. *Journal of Japanese Linguistics* 13: 91–116.

Uyeno, Tazuko. 1971. "A Study of Japanese Modality: A Performative Analysis of Sentence Particles." Ph.D. dissertation. University of Michigan.

Vendler, Zeno. 1957. Verbs and Times. In *Linguistics in Philosophy*, edited by Zeno Vendler. Ithaca, N.Y.: Cornell University Press.

Verkuyl, Henk. 1993. *A Theory of Aspectuality: The Interaction Between Temporal and Atemporal Structure*. New York: Press Syndicate of the University of Cambridge.

White, Alan. 1979. What Might Have Been. *American Philosophical Quarterly* 4: 102–29.

Wright, Georg Henric von. 1951. Deontic Logic. *Mind* 60: 1–15.

Yamada Yoshio. 1951. *Nihon Bunpō-gaku Gairon*. Tokyo: Hōbun-kan.

Yoshikawa Taketoki. 1976. *Gendai Nihongo-Dōshi no Asupekuto no Kenkyū. Nihongodōshi no Asupekuto*. Tokyo: Mugi Shobō.

Index

Note: Page numbers in italic refer to illustrations or graphs.

O

P

About the Author

Yuki Johnson received her Ph.D. in Japanese linguistics from the University of Minnesota. She served many years as Director of the Japanese Language Program at the University of Michigan and then at the University of British Columbia, and is now an Associate Professor of Japanese Linguistics at the University of Toronto. Currently she is completing a manuscript on Japanese linguistics titled "Fundamentals of Japanese Grammar."